Lust and Love
Is it More Than Chemistry?

For Jeremy

Lust and Love
Is it More Than Chemistry?

Gabriele and Rolf Froböse

RSCPublishing

Originally published in the German language by WILEY-VCH Verlag GmbH & Co KGaA, Boschstraße 12, D-69469 Weinheim, Federal Republic of Germany, under the title "Froböse, Froböse: Lust und Liebe – alles nur Chemie?". Copyright 2004 by WILEY-VCH Verlag GmbH & Co KGaA.

Translated and edited by Michael Gross, Oxford

ISBN-10: 0-85404-867-7
ISBN-13: 978-0-85404-867-0

A catalogue record for this book is available from the British Library

Published by The Royal Society of Chemistry,
Thomas Graham House, Science Park, Milton Road,
Cambridge CB4 0WF, UK

Registered Charity Number 207890

For further information see our web site at www.rsc.org

Typeset by Macmillan India Ltd, Bangalore, India
Printed by Henry Ling Ltd, Dorchester, Dorset, UK

Preface

Love presumably is as old as mankind. Accordingly, it is a thread that runs through innumerable books, from antiquity through to modern literature. The situation is similar in the visual arts, where the "topic No. 1" also takes centre stage. Desire and passion have also inspired music, from mediaeval minnesong, via French chanson and the Beatles, through to today's popular music.

Nevertheless, love is also a phenomenon that is poorly understood. It is described in terms such as "undescribably beautiful," "mysterious," or "exciting." But what is it really all about?

"Love is just a word," the Austrian author Johannes Mario Simmel said, possibly reflecting his personal experience. Slightly more promising, Henry Miller called it a "boogie-woogie of emotions."

The "poems in prose" of the German author Margot Bickel relate the longing of all human beings for happiness, peace, and togetherness. On the nature of love she writes:

"What is love?
Perhaps, silent understanding,
patient questioning, understanding perseverance,
tender togetherness, reliable standing up for each other,
common search, conciliatory arguments,
to be there and stay there,
and now and again roses and a kiss?
What is love?
I only know facets
which give me an idea
of the humility and the majesty of love.
What is love?
Let us love and find the answer."

In contrast to Margot Bickel who withholds the answer, fellow German author Andreas Mäckler provides many. In his book "Was ist Liebe . . . ? 1001 Zitate geben 1001 Antworten," he provides an overwhelming variety of statements, hypotheses, claims and affirmations ranking around the topic of love.

"Love is chemistry!" When the authors make this bold claim it's not a question of adding a 1002nd thesis about love to the catalogue of interpretations. While this definition does not flow from the pen of a philosopher or literary genius, it is not meant to be a provocative statement floating in mid-air either. Rather, it is the summary of modern interdisciplinary research involving chemists, medical experts, brain researchers, hormone specialists, and biochemists. Together they have succeeded in gradually understanding the human life processes and in interpreting the underlying functions in terms of chemical reactions.

With this, we don't want to degrade life to "pure chemistry." On the other hand, we should emphasize that chemistry is not a human invention, but that everything in Nature is based on it. This is true for the dust of the Sahara desert as well as for the plankton of the oceans and for the rock crystal as well as for plants, animals and humans. Even if we enjoy a delicious meal, experience joy, or fall in love, chemistry is pulling the strings behind the scenes.

We invite you to join us and our fictional couple Bianca and Michael on a fantastic journey through the chemistry of the senses, which governs, love, desire, and passion, and thus makes life worth living.

Gabriele und Dr. Rolf Froböse
Wasserburg am Inn (Germany), December 2005

Contents

Chapter 1 Mysteries of the Emotional Rollercoaster   1
1 A Day in the Life of Bianca and Michael   1
2 Why Our Brain Produces "Butterflies"   2
  2.1 Mapping the Brain is Like Decoding the Genome   5
  2.2 The Universe Inside Bianca's and Michael's Heads   6
  2.3 Think Before You Act! – Order Refused, Says the
      Spinal Cord   10
3 Michael's Brain, a Dream Super Computer   12
  3.1 Is Our Brain a Parallel Computer?   12
  3.2 Right or Left? Our Split Brain has to Make a
      Decision   14
  3.3 Kissers Turn Right (Probably)   15
  3.4 Can Thoughts be Measured?   17
  3.5 A Formula 1 Racing Car in Michael's Brain   18
4 The Cell as a Chemical Reactor   19
  4.1 The First Encounter With Bianca: Michael's Ion
      Channels "Remember"   20
  4.2 Michael has Localised Bianca: The "Chemistry of
      the Moment"   20
  4.3 Michael's and Bianca's Emotions are Mainly Hap-
      pening at the Front of the Right Hemisphere   23
  4.4 "Squids" Made of Niobium as a Compass of our
      Thoughts   24
5 Outbursts of Emotion Set Molecular "Submarines"
  Going   25
  5.1 Michael's Love Triggers an Avalanche of Signals
      in the Brain   27
6 Neurotransmitters – Guardians of Emotions and
  Sleep   28
  6.1 Is Michael's Brain Squandering Energy?   30
  6.2 Bianca Likes Glucose – Her Brain Too   31

Chapter 2 Two Networks of Nerves   33
  1 Hugs and Heartbeats   33
    1.1 Control Over Sex and Senses   33
    1.2 Team Work   34
    1.3 No Erection without the Cooperation of Vegetative and Somatic System   35

Chapter 3 Signals of Love   36
  1 Love at First Sight?   36
    1.1 Shy Men Avoid Eye Contact–Women Look for it   37
  2 Anatomy of a Flirt   38
  3 The American Kama Sutra: 103 Pulling Tactics   39

Chapter 4 Tied Up with a Double Helix   41
  1 All in Our Genes?   41
  2 From Primordial Soup to DNA   42
  3 Chemistry: The Seed of Life, Mind, and Emotions   44
  4 Genes are Like Men   45
    4.1 Why does Michael have Dark Hair? Sugar and Phosphate Molecules Provide the Answer   47
  5 Inherited Genes: A Surprise Present   47

Chapter 5 Hormones – The Body's Snailmail   49
  1 Blame it on the Adrenaline   49
    1.1 Adrenaline and Co.: the Secret Rulers of Our Emotional World   49
    1.2 Hormone Deficiency in the Brain can Trigger a Rollercoaster of Emotions   52
    1.3 Checks and Balances   54
    1.4 Exultation and Melancholy   55

Chapter 6 Estrogen and Testosterone – Driving Our Emotional Lives   60
  1 The Secret Entanglement of Body and Soul   60
  2 Bianca Knows the Story but can't Fight her Feelings   60
    2.1 During Puberty, the Countdown of the Body's Chemistry Begins   62
    2.2 Men are Much Less Complicated, or are They?   62
    2.3 Testosterone – A Versatile Hormone   64
  3 Love is the Best Medicine   66
    3.1 Domestic Disputes Weaken the Immune System   67

Chapter 7 Oxytocin – the "Amuse-Gueule" Among Hormones   68
 1 When Michael Caressed Bianca, her "Chemical Factory" Produced Oxytocin   68
  1.1 Keen on Cuddling after Orgasm? It's all in the Chemistry!   69
  1.2 From Moscow with Love: Oxytocin and the Pavlov Effect   70
  1.3 Why Women Pick up the Phone When They are Stressed   72
  1.4 Small Molecule, Big Effect   73
  1.5 Men and Women, Be Tolerant!   75
  1.6 Social Behaviour Benefits, Too.   76
  1.7 Oxytocin and Vasopressin – Chemical Compass Needles for Partnership and Fidelity   76

Chapter 8 Dopamine – Casanova's Double-Edged "Secret Weapon"   78
 1 A Hormone that makes us Euphoric   78
  1.1 Excess Amounts of Dopamine can Lead to Pathological Addiction to Love   79
  1.2 Male Rats Show a Pronounced "Coolidge Effect"   79

Chapter 9 Serotonin – the Happy Messenger in the Blood   82
 1 A Moody Messenger   82
  1.1 Sometimes Michael Craves Chocolate   83
  1.2 Pathological Love – Linked to Serotonin Deficiency   83
  1.3 Microparanoia – The Love Sickness   84

Chapter 10 Phenylethylamine – the Stuff that Makes the Soul Jubilate   85
 1 Welcome to the Rollercoaster of Emotions   85
  1.1 The Elevator to Cloud 9 Smells of Fish   85

Chapter 11 The Chemistry of Birth Control   87
 1 How about the Pill?   87
  1.1 Birth Control – Almost a Never Ending Story   87
  1.2 Just Crouch Down and have a Good Sneeze   88
  1.3 A Big Leap for the Biology of Small Eggs   89
  1.4 Why is There No Pill for the Male?   92

Chapter 12 Menopause: When the Hormone Supply Falters   93
 1 Will You Always Love Me?   93
  1.1 Is Hormone Replacement Therapy the Way Out?   94

1.2 Drugs from Nature's Treasure Trove   95
1.3 Not only Women   96

Chapter 13 Endogenous Opiates – the Chemistry of Euphoria   100
1 It's not only the Adrenaline that Gets Michael and
   Bianca Started   100
   1.1 Is Our Brain a Poppy Flower?   101
   1.2 Researchers Go the Whole Hog   102
   1.3 Long Time no See?   103
2 Endorphins – An Emergency Centre of the Body?   105
   2.1 Mice Turn into Little Cowards   107
   2.2 Faith is the Best Medicine – Endorphins and The
        Placebo Effect   107
3 Endorphins in Pain Research   108
   3.1 Opiates Linked to Near Death Experience?   110
   3.2 Hugging the Pain Away   110
   3.3 Immediately After Birth, Mother and Baby are full
        of Opiates   111
   3.4 Breastfeeding Makes Babies High   111
4 Naloxone – Opponent of the Opiates   112
   4.1 Without Endorphins, The World of Intensive
        Emotions Remains Locked   113
5 Natural Opiates   113
6 Synthetic Opioids   114
   6.1 Synthetic Opiates were Meant to Knock Out
        Terrorists   115
7 Are Our Opiate Receptors a Blessing or a Curse of
   Nature?   116

**Chapter 14 Chemistry for the Eye of the Beholder – Lipsticks through the
Ages**   117
1 Bianca's Red Lips get Michael all Excited   117
   1.1 In the Ice Age, Ochre was all the Rage   117
   1.2 A Magic Wand of Cosmetics – Thanks to
        Chemistry   118

Chapter 15 Secret Scents of Seduction   120
1 Come on let's Go   120
   1.1 From the "Dialogue with the Gods" to Modern
        Perfume   120
   1.2 Top Notes, Middle Notes, and Base Notes: An
        Excursion into the Chemistry of Fragrances   123

1.3 A Frenchman had a Nose for it 124
1.4 Animal Instincts 125
1.5 Take a Deep Breath – How Our Sense of Smell Works 125
1.6 Is the Gradual Loss of Our Sense of Smell an Error of Evolution? 126
1.7 Show Your True Colours and I Tell You Which Fragrance You Prefer 127

Chapter 16 Pheromones – Words in the Dialogue of Fragrances 130
1 Do Pheromones make Humans Horny, Too? 131
1.1 A Relic from Earlier Days of Evolution? 133
1.2 The Truffle Pig in us Decides Whether We Get on with Each Other 135
1.3 Men are Smellier Than Women 135
1.4 Napoleon to His Wife: "Don't Wash, Will Arrive in Three Days." 137
1.5 Copulins – the "Chemical Weapons" of Women 137
1.6 Artificial Pheromones in Perfumes make Men Keen to Cuddle 138
1.7 "I Don't Like Your Smell, Because Our Genes are Too Similar" 140

Chapter 17 The ABC of Aphrodisiacs 142
1 A Little Help from My Friends 142
1.1 E is for Eggs 143
1.2 G is for Gelée Royal and Ginseng 143
1.3 L is for Licorice 145
1.4 O is for Oysters 145
1.5 P is for Papaverine Injections 146
1.6 R is for Rhinoceros Horn 148
1.7 S is for Spanish Fly 149
1.8 T is for Tiririca 150
1.9 V is for Vitamin E 151
1.10 Y is for yohimbine 152

Chapter 18 Viagra & Co 153
1 A Growing Problem 153
1.1 The Discovery of Viagra 154
1.2 The Penis as a Book-Keeper 154
1.3 Viagra – Just What the Doctor Ordered? 155
1.4 Potent Drugs Fight Impotence 156

Chapter 19 Epilogue: Returning from the Airport 157
    1   The Important Thing is That The Chemistry Works –
       The Rest Remains a Miracle 157

Further Reading 159

Subject Index 163

CHAPTER 1

Mysteries of the Emotional Rollercoaster

Lovers close their eyes when they kiss,
because they want to see with their hearts.
Daphne du Maurier (Writer, 1907–1989)

1 A DAY IN THE LIFE OF BIANCA AND MICHAEL

Never before has Michael arrived at an airport so early. Today is a very special day. With a quick look at the monitor he finds out that the transatlantic flight he is waiting for will arrive at gate B14 in an hour.

On board that flight is Bianca, to whom he has been happily engaged for six months. Bianca, a medical student, has spent a large part of the summer holidays with relatives in the US, where she had an internship at a hospital, and she must have had many interesting experiences. For Michael, however, an IT engineer for a European electronics company, the separation appeared like an eternity. While he always kept himself busy, he hates to think of the long weekends spent alone.

"Well, in that case, I could have taken it slowly," he mused by himself, "but never mind, better to be early at the airport than late." The thought of getting stuck in traffic, while Bianca, with her suitcases in hand, might be looking for him in vain, makes him uncomfortable.

Strolling through the arrivals area, Michael checks one more time which exit Bianca will be taking coming from B14. He realises how his inner tension relaxes gradually and makes way for deeply felt joy. To pass the time, he takes a seat in the cafe called "Zeppelin," allowing him a direct view at the monitor displaying the arrivals.

"I'll have a chicken sandwich and a cup of coffee," he tells the waiter and reaches for a newspaper. He scans the headlines, but only skips diagonally through the articles. He finds it hard to concentrate today. Only a feature about San Francisco holds his attention. "Her relatives

1

live in Monterey," he recalls. "Knowing her, she hasn't been content with seeing the photos of Golden Gate Bridge."

While Michael is busy eating, the monitor updates the "arrivals" information. The plane arrives a little earlier after all and will land in a few minutes, he realises. Nervously, he folds up the paper and signals the waiter. After paying the bill, he walks straight to the exit.

Excitedly, Michael observes the sliding door opening and closing at short intervals. Tanned tourists heavily loaded with suitcases and bags struggle through the waiting crowd. Businessmen with briefcases hurry past him, while a Japanese man is upset, apparently looking for somebody, and a group of three Arabs enjoy their small talk. Michael notices these figures only marginally, as he would the extras in a movie.

Suddenly his expression relaxes – Bianca comes through the door. She spots him immediately, leaves the trolley with the suitcase behind and runs towards him. Speechlessly, they sink into each other's arms. When they kiss, Bianca has tears of joy in her eyes. Michael, noticing the familiar smell of her body, has only one thought on his mind: "We belong together!"

This short scene from the life of two young people should sound familiar to many of us. The unrest, tension, excitement, combined with longing, and then the unlimited joy after the encounter – who hasn't experienced this rollercoaster of emotions in similar situations?

2   WHY OUR BRAIN PRODUCES "BUTTERFLIES"

Even though those two may be under the impression that the centre of their love is located in the heart, in reality it is only the brain that is responsible for the heartbeats and the "butterflies" in the tummy. "We do not think with the heart, but with the brain," as the Greek physician Hippocrates, who lived on the island of Kos, stated around 400 BC – but he was way ahead of his time (Figure 1). Even though the organ which the Greek called "en kephale" (located in the head) had fascinated humankind from the beginnings, it was a long way to the understanding that only the brain is the source of our thoughts, feelings, sensations, and ultimately our consciousness.

Even our ancestors in prehistoric times must have been wondering about the source and location of consciousness. Thus, people in ancient cultures saw the head as the dwelling of evil spirits. As we know from skeletons found, some people of that era had holes carved in their skulls – apparently with the aim of curing maladies like "obsession," although the success must have been dubious.

Figure 1 *Hippocrates*

Greek anatomists like Anaxagoras looked for the location of the mind in the human body and believed that the cavities in the brain contained the fluid which represented the breath of the mind. Around 500 BC, the Greek Alcmaeon of Croton conducted sections of animals and found out that nerves connect the sensory organs to the brain. He concluded that the brain contains the centre of sensory perception and thinking. However, he considered the brain to be a gland that secretes thoughts like the tear glands produce tears.

The Ancient Egyptians also connected human thought processes with the brain. Herophilus (335 BC) and Erasistratos (300 BC) were the first to break the taboo against dissections of human bodies. They found that people who had certain nerve paths cut, were no longer able to see. Thus they developed the concept of an interconnected system, of which the brain was the centre. To them, the brain was the seat of the soul and the central command of all thought processes.

The Roman physician Claudius Galenus had the opportunity to gain insights from numerous injured gladiators. With this work, he helped to establish the concept developed by the Egyptians that the brain is the centre of human thinking and memory. Aristotle, in contrast, held very different views. Unlike Hippocrates, he insisted on the view – still favoured by the romantically inclined – that human beings think with their hearts.

Eventually, the chamber model of Anaxagoras, which was improved over the centuries, remained victorious. Mediaeval philosophers turned it into a very vivid model, in which the first chamber of the brain serves perception and insight. The second chamber, according to the model, is for knowledge and judgement, while the third chamber is in charge of recording the results of the previous two chambers.

Figure 2   *Leonardo da Vinci*

LEONARDO DA VINCI: ARTIST AND RESEARCHER

When Leonardo da Vinci was born in 1452, Italy was about to leave the Dark Ages behind at a fast pace. Italy, and Florence in particular were at the centre of the intellectual life newly awakened during the Renaissance. This historical development, which had its beginnings in the learned circles of the humanist writers, was clearly connected with the progress made in the sciences, the changes in the clerical realm, and the emergence of economic structures.

Leonardo da Vinci (Figure 2) was the son of a respected notary. The father recognised the unusual talent of his son quite early on and supported him with all means at his disposal. Thus, young Leonardo at the age of 15 arrived at the atelier of the Florence master Verrocchio, and by 1472, at the age of just 20, he had already made his name among the painters of the city.

From about 1500, Leonardo da Vinci dedicated himself mainly to technical and scientific studies. In countless, very precise drawings of muscles, bones, and brains he tried to track down the laws of life and combine them in a cosmology including all phenomena of nature.

Even around 1490, Renaissance's all-round genius, Leonardo da Vinci, drafted a preliminary "map of the mind," which associated different mental functions with various areas of the brain divided in three parts.

Although Leonardo's sketches of the brain are today only of historical interest, making place for a much more refined picture of the brain and its functions, our most intimate organ has yet to reveal many of its secrets. Even today, many aspects of brain function appear as blank spots on the map of scientific knowledge.

The French philosopher René Descartes (1596–1650) had a much more technical view, comparing the brain with a kind of machine. He imagined that a substance contained in the windings of the brain, which he called "pneuma," is put under pressure by the excitation arising from the sensory organs and directed into the tube-like nerves by the pineal gland. Thus, the pneuma would travel to the muscles and make them move.

Franz Josef Gall (1758–1828) stirred up his contemporaries by asserting that certain actions of the brain can be felt through the skull. But only Paul Broca (1824–1880) and Carl Wernicke (1848–1905) provided scientific evidence showing that brain functions can be assigned to specific regions. For this purpose, the researchers had investigated a number of patients with language disorders. Between 1900 and 1920, Cecile and Oskar Vogt, along with Korbinian Brodman led this work to its logical consequence and drew the first detailed "architectural" maps of the cortex.

2.1 Mapping the Brain is Like Decoding the Genome

While the earlier thinkers believed that complex processes like learning or memory can each be confined to one particular area of the brain, today's scientists assume that each brain activity involves various groups of cells that may be distant in space but are linked by nerve fibres. Researchers led by Karl Zilles at the Jülich Research Centre (Germany) have committed themselves to the task of localising these nodes and networks. Their ultimate goal – the complete mapping of all brain functions – is an extremely ambitious one and could be compared to the decoding of the human genome. It is already becoming obvious that the results of this research will throw up a multitude of new questions, which will keep generations of scientists busy.

It is beyond doubt that the brain is our central command which governs all body functions. This applies not only to simple behavioural patterns like eating, sleeping, drinking and heat regulation, but includes the more highly developed abilities of the human mind such as its gift for culture, music, art, science, and language. But it was only recently that researchers obtained insights into the molecular processes in the brain and decoded the first building blocks and processes of a hitherto unknown chemistry which controls all our thought processes – be they conscious or unconscious – and thus our entire emotional world including love. When Bianca and Michael ran towards each other at the airport, when they hugged and cuddled each other, these events triggered a whole cascade of chemical reactions in their brains.

2.2   The Universe Inside Bianca's and Michael's Heads

And yet, one should not imagine the brains of our protagonists as a simple chemical reactor. Because the brain would be completely useless if it wasn't connected with the entire human body with an unimaginably complex network of wires. A mesh of around 380,000 nerve fibres, which would span the distance to the Moon if they were aligned end to end, ensures the smooth flow of information between the central command and all other areas of the human body.

It may sound unbelievable, but the hardware in our heads consists of around 100 billion nerve cells, and is thus comparable to the number of stars in the Milky Way. If we were to calculate the number of connections between these cells that would be possible in theory, the result would be absolutely mindboggling, as there are more possibilities than there are atoms in the entire Universe!

Werner Stangl, a professor at the institute for pedagogics and psychology of the Johannes Kepler University of Linz, Austria, goes even further and visualizes this unimaginably large number thus: "If the brain contains at least 15 billion cells, their connective possibilities could in theory store 210 billion pieces of information. If we wanted to write down this number at the rate of one figure per second, it would take us 90 years."

This unique architecture allows the brain to do more than simply represent the information it acquires. Unlike a camera or a tape recorder, it has ingenious ways of data reduction. In other words, the brain separates unnecessary junk data by interpreting the signals recorded from the outside world within fractions of a second, and summarising them to form a personalised world. While Michael was waiting for Bianca at the airport, his brain received an incredible one million times more information than his consciousness processed.

THE AGEING BRAIN LOSES ONLY A SMALL NUMBER OF CELLS

According to Prof. Werner Stangl, we lose between 1,000 and 10,000 brain cells every day. "Even if we assume that a person loses 10,000 of their initial reservoir of 15 billion cells every day, they would have to reach an age of 410 years in order to lose just ten percent of their brain," he calculates. This calculation clarifies that the capacity of the brain cannot be the limiting factor if the memory abilities decline with age. The reasons for the decay are usually found in the lack of training. When people are no longer challenged by their environment and their working life, when they no longer have to learn and the intellectual requirements are reduced, they must take measures themselves and train their brain. Only intellectual activity can ensure that new brain patterns and structures are formed. This way, the thinking and memory abilities can not only be conserved, they can even be enhanced in old age, says Stangl.

In order to better understand this remarkable process, which no computer can match, we shall have a closer look at the human brain. If Bianca and Michael represent the statistical average, her brain weighs around 1245 grams, and his 1375 grams. The largest part is the cerebrum, which has the size of a grapefruit. It consists of two different halves, the left and right hemispheres, which are, among other things, in charge of the physical functions of the body half located on the opposite side. The hemispheres are covered by the multiple folds and creases of the cerebral cortex. The cortex enables us to organise, remember, understand, to communicate and be creative, to invent and value things. The most complicated and remarkable part of the brain however, is the pea-sized hypothalamus, the "brain" of the brain, so to speak. It controls basic needs like eating, drinking, sleep, but also body temperature, pulse rate, hormones and sexuality. By a combination of electrical and chemical messages, the hypothalamus also controls the pituitary gland. The latter is the most important gland of our central control station and controls our body with the help of hormones – chemical messengers which reach their target cells via the blood stream (see Chapter 5).

OUR BRAIN AT A GLANCE

Roughly speaking, the human brain can be divided in five areas: brain stem, cerebellum, limbic system, cerebrum, and the lobes of the cortex.

The Brain Stem

This is the oldest part of the brain. It developed more than 500 million years ago. As it resembles the complete brain of a reptile, it is also called the reptilian brain. Our animal ancestors already had this brain, therefore it is in charge of all fundamental functions of life:

movements, hunting, grooming, territorial marking, rites, mating, habituation. The reptilian brain also controls vital functions like breathing and pulse frequency. As it conserves ancestral habits and behaviours in practically unchangeable ways, it has only a limited learning ability, but it can give us the sensation of routine and safety. In contrast, the reptilian brain does not know any emotions. Therefore, the Swiss psychoanalyst C.G. Jung (1875–1961) claimed that certain archetypical behaviours from the earliest times of humankind are based in the reptilian brain.

The Cerebellum

This part is located at the back of the brain stem, below the cerebrum, in the lower rear part of the skull (see Figure 3). Like the cerebrum, it consists of two hemispheres. Its size has roughly tripled during the last one million years of human evolution. The cerebellum is mainly responsible for the correct execution of body movements, and for orientation in space. It also serves to store memories and to carry out simple, acquired functions. The ability to learn new movements and recall them "automatically" at a later time is also located here. The cerebellum stores all movement sequences that we learn, from throwing a ball through to playing the piano.

When we walk, run, or grasp an object, everything seems to happen automatically. But in fact, every movement requires a lot of coordination. Only when we are immersed in a new kind of environment do we realise how much work the brain has to do. Astronauts exposed to zero gravity, for example, have to relearn even simple movements.

The Limbic System or Mammalian Brain

The limbic system is a more recent development in evolution. This area which is most highly developed in mammals can look back on 200–300 million years of evolutionary history. It is involved in the control of body temperature, blood pressure, pulse frequency, and glucose levels in the blood. Furthermore, it has an important part to play in vital emotional responses. Laughing and crying, playfulness and sexuality, euphoria and depression are based here. All information bound to be stored in the long-term memory first passes through this part of the brain. Rational cognition and emotions meet here.

The key elements of this area are the hypothalamus and the pituitary gland. Although it is only pea-sized, the hypothalamus controls important functions like eating, drinking, sleep, wakefulness, body temperature, and many others. This control mechanism is based on a multitude of electrical and chemical messages through which the hypothalamus controls the pituitary gland.

The Cerebrum

The largest part of the human brain is the cerebrum. It accounts for about 85% of the entire brain mass and includes the highly developed outer layer, the cerebral cortex. The cortex contains the centres responsible for movements, speech, seeing and hearing.

Furthermore, our cerebrum is the seat of consciousness, will, intelligence, memory, and learning ability. It consists of two strongly creased hemispheres, separated by a deep incision. The left hemisphere controls the right half of the body and vice versa. Thus, nerve cells in the left motor field are activated if the right hand is touched. Both parts are connected via a thick strand of nerves, known as the corpus collosum, implying that they frequently exchange information. Different kinds of tasks that humans are confronted with are distributed unequally between the hemispheres. For example, the sense of time and linguistic ability are mainly on the left, musicality and rhythm on the right. Information processing is handled differently as well: on the left, it happens serially (one thing after the other), while on the right it is more parallel (*i.e.* many things at one time). Damage to one half of the brain often leads to the suppression of all sensory and motor functions of the opposite side of the body. This is often observed after a stroke.

For most people, the left half of the brain is dominant. This is why there are more right- than left-handers, even though cultural influences also contribute to this.

The Cortical Lobes

In both halves, the cortex is divided into four areas, known as lobes. Of these, the frontal lobes are necessary mostly for planning, decision-making, and target-oriented behaviour. The apical lobes represent the body – they receive the sensory information. A part of the parietal lobes is responsible for vision and therefore known as the visual cortex. The temporal lobes seem to have several important functions. Among them are hearing, sensory consciousness, and memory. The cortex is the seat of the sensory perceptions and their connection to the movement apparatus and the intellectual abilities. For humans it is the organ most important for survival, as it generates crucial abilities like cognition, thinking, combining, memory –

the prerequisites for everything we call learning. There is a very complex interplay between incoming information, its processing and archiving, and the transmission of commands to the organs of movement.

2.3   Think Before You Act! – Order Refused, Says the Spinal Cord

Together with the brain, the spinal cord represents the central nervous system, or CNS. The spinal cord serves as a cable for communications, enabling messages from the brain to be transmitted to the rest of the body at high speed. However, it also acts independently in controlling a number of reflexes.

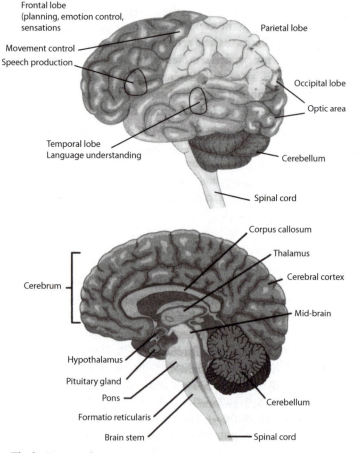

Figure 3   *The brain at a glance*

This will remind many readers of the proverbial knee-jerk, which a neurologist can trigger with a tap to the tendon just under the kneecap. The tapping causes a brief extension of the extensor muscle, which in turn leads to its contraction. Many who have felt the hammer of a physician will wonder what this reflex is for. Nature very wisely gave us this reflex when we started to walk upright, because without it we would not be able to stand straight without our knees buckling up every now and again. For the neurologist, in turn, testing the knee-jerk is an important diagnostic tool, as its absence could point to a serious disorder of the CNS.

"Think before you act" – may be a useful instruction generally, but it does not apply to reflexes. They represent instant measures, which are immediately put into action by the spinal cord without consultation of the brain. For example, when you accidentally touch a hot object and your hand withdraws automatically with lightning speed, although pain and realisation of the danger only register later. This withdrawal reflex is completely unconscious. With the help of receptors in the hand, switchable neurons in the spinal cord, and motor neurons leading to the upper arm muscles, the worst outcome can be avoided.

A cross section of the spinal cord appears as a round disk the width of a finger. Its core consists of the so-called grey matter, shaped like a butterfly. It is composed of closely packed nerve cell bodies, and its outer coat consists of nerve cell fibres, the so-called white matter.

Depending on the overall height of a person, the spinal cord can measure up to around 45 centimetres, but on average it weighs only 25 grams. It begins at the extended marrow of the brain and runs through the channel of the vertebrae through to the height of the second lumbar vertebra. At regular intervals, pairs of nerve roots branch out from the marrow on both sides. The dorsal root, *i.e.* the part of the cord oriented towards the back, carries impulses from everywhere in the body to the grey matter of the spinal cord. Conversely, the ventral (*i.e.* oriented towards the front) nerve root, the motor neuron, directs impulses towards the muscles. Only a few millimetres beyond the point where they leave the spinal cord, the roots recombine to form the so-called spinal nerves. These emerge from the vertebral channel via the intervertebral foramen, *i.e.* the holes between the vertebrae.

In the neck and lumbar region, the spinal cord becomes much thicker. In these areas, many nerve fibres emerge, leading towards the arms and legs, respectively. Even though the spinal cord itself ends at the height of the second lumbar vertebra, the nerve fibres from the lower part of the spinal cord run further down within the vertebral channel. They are combined to a thick bundle of fibres, which by and by, indvidual nerve fibres emerge between the vertebrae. This bundle is reminiscent of a horse's tail and has therefore been named with the Latin word for it, "cauda equina."

3 MICHAEL'S BRAIN, A DREAM SUPER COMPUTER

What happened in Michael's head while he was waiting for Bianca at the airport? Hundreds of people, the restaurant, check-in desks, monitors, announcements... the entire hectic atmosphere of a busy airport confronted him with an avalanche of optical information, which was duly recorded by his eyes and triggered light-sensitive nerve cells, which in turn converted it into electrical impulses. Then the information arrived at the visual cortex, where it was subjected to a thorough analysis. During the selection process, Michael's visual cortex packaged the electrical impulses into concrete information, which is transmitted to the temporal lobe, where for the first time, it may or may not be recorded by his memory.

But the odyssey of information flow did not end there. From the temporal lobe it proceeded into the depths of his brain, arriving first in the structures of the frontal lobes. There, a further filtering of the data took place, after which the selected and evaluated information was cleared for forwarding to the entire cerebral cortex. Only this continuous to and thro of thousands of information details in the "circuits" of Michael's brain enabled him to crystallise his personal world out of the information overload and to locate Bianca in the crowd of strangers. At the same time, his motor cortex enabled him to make the unmistakable signs of recognition (*e.g.* the joy in his face) visible.

3.1 Is Our Brain a Parallel Computer?

Thus, our brain is a gigantic network of more than 100,000 km length of cable. Like electrical cables, nerves conduct electricity. If a nerve cell is triggered by an arriving stimulus, its state changes rapidly: it will either be excited or inhibited. When a cell gets excited, there will be a chain reaction, in which molecular messengers will excite the neighbouring cells and their neighbours as well.

Ernst Pöppel of the institute for medical psychology at the university of Munich suspects that every nerve cell stays continously in contact with 10,000 other nerve cells. This means firstly that 10,000 cells are influenced by a single nerve cell (divergence principle) and it also means that each nerve cell is influenced by 10,000 others (convergence principle). These contacts can be stimulating (excitation) or inhibiting.

Excitation and inhibition are signalled by different chemical messengers, known as neurotransmitters.

"Although there are lots of nerve cells in the brain, its working mechanisms are also characterised by the 'strong law of small numbers,' which leads to the functional proximity of nerve cells. Each nerve cell is no more than four steps away from any other nerve cell in the brain. (This corresponds to the observations made in the recent studies of "small-world networks" in nature and technology, such as the internet.) Translated into the language of IT, this functional proximity implies that the brain commands "massive parallelism," as everything is well connected to everything else.

Pöppel goes one step further and draws the far-reaching conclusion that "a simulation of human thinking is still extremely remote." This is because the brain has a fundamentally different architecture from all computers, and the principles of its data processing are fundamentally different from conventional algorithms and their implementation in computer programs. Therefore, says Pöppel, the simulation or explicit modelling of human thinking, perception, feeling, decisions, remembering, or actions has so far remained the domain of science fiction.

Solely from the architecture of the brain the scientist draws the conclusion that perception without simultaneous remembering and emotional evaluation (or conversely, remembering without emotional evaluation and perception) or an emotion without a reference to memory and a perceptual representation is impossible.

Only in retrospective reflexion can we discover or invent seemingly independent phenomenologies, for which we use concepts like perception, memory, or emotion. In the ongoing flow of experience and in immediate actions, this separation does not exist, says Pöppel.

There is another finding of modern neuroscience that deserves special attention because of its potential importance for other areas. Newborn babies are equipped with a lavish supply of possible connections between nerve cells. According to experts like Pöppel, however, this genetically predetermined potential becomes effective only if the various associations of nerve cells and their genetically offered connections are in fact being used. Only by the use of the local information processing capacity this potential becomes a resource which in the long term influences the behaviour and indeed the life of a person.

It is the functional use of these connections that creates the detailed structure of the brain. What isn't used will be switched off permanently. Thus, potential connections between nerve cells are not necessarily maintained throughout life. Effortless learning – for example learning languages in early childhood – is no longer possible later in life, as the

Figure 4   *Optical illusion*

learning processes will then have to run within the established brain structures. The young child has a remarkable gift for the correct phonetics, and for remembering language patterns. From around age 10, the cognitive thought processes are more developed, such that language learning is increasingly guided by logical connections, *e.g.* grammar. Thus, the brain is not a passive filter, but it constructs its own world. "The brain possesses a creative force," Pöppel confirms.

Evidence for this is provided by the well-known optical illusions created by pictures with a double meaning, in which one can see one of two things, *e.g.* either an old lady or a young lady. One can never see both versions at the same time, but one can switch between the two visual modes intentionally (Figure 4). Thus the brain can create different interpretations of one and the same visual input.

3.2   Right or Left? Our Split Brain has to Make a Decision

As we mentioned above, the cerebrum is the youngest part of the human brain in terms of evolutionary history. Many scientific studies have shown that there is a clear division of labour between its two halves: The left hemisphere is associated with functions like logic, language, and analytical thinking, while the right hemisphere is credited with musicality, creativity, and spatial representation. Furthermore, each half controls the movements of the opposite side of the body.

"Based on this division of functions, there have been demands for a learning style that involves more contributions from the right half of the cerebral cortex," Werner Stangl explains. However, it has also been shown that such demands must be taken with a pinch of salt. While the hemispheres are indeed specialised, there is no strict separation of their functions.

In fact, the right hemisphere, which commands an extensive dictionary, can also fulfil tasks in language generation. On the other hand, the left hemisphere is also involved in processing music. Thus, drawing conclusions about learning from the specialisation of the hemispheres can be problematic. Studies of the blood circulation in the brain during the intake of emotionally charged information have confirmed this cautionary note.

For a long time, researchers have wondered why humans tend to prefer the right hand side, even though one might have expected a 50:50 distribution of right and left handedness, for example. According to Stangl, even unborn babies in the womb have a preference for the right side, which is also observed in newborns. However, this effect disappears at the age of three to six months. As the asymmetrical preferences of hands, feet, eyes and ears become prominent only later in life, it remains unclear whether they are related to the baby's tendency to turn in a certain direction.

3.3 Kissers Turn Right (Probably)

Most people have a preference for the right eye, ear, foot and hand, with a right:left ration of around 2:1. Birds, too prefer the right eye, thus they prefer to turn their head to the right even as embryos, as most vertebrates do. Thus, even before hatching, the right eye gets most of the light stimulation. Researchers at Bochum, Germany, have established that this effect changes the young bird brain in asymmetric ways, and that this asymmetry leads to further right/left differences in perception and cognitive processes.

To prove that adults have a preference to turn their heads to a given side, one has to observe them in a situation where they decide to turn spontaneously and without pressure from outside. The biopsychologist Onur Güntürkün, of the Ruhr University at Bochum, realised that he could investigate such a situation if he played voyeur for a while.

"I had this idea to observe couples kissing," explains Güntürkün. No sooner said than done. During two-and-a-half years he used waiting times at airports and railway stations, as well as stays on beaches and in parks in Germany, the US, and Turkey to collect data for his study. During this time, he examined 124 kisses of couples between around 13 and 70 years of age, taking into account only one kiss per couple, and in the case of prolonged activity, only the first kiss he observed. In order to qualify for his study, each kiss had to satisfy four criteria:

- It had to involve lip-to-lip contact,
- the kissers had to stand opposite each other,

- neither of them was allowed to hold any objects in their hand (as this might trigger a preference for one side or the other), and
- there had to be an unequivocally observable turn of the head.

The results showed that 80 out of the 124 qualifying kisses were carried out with heads turned to the right hand side. From this, Güntürkün concluded that the right bias of the head is retained in adulthood and that presumably the other asymmetries of perception and action follow from it.

Intriguingly, the voyeur in the service of science never observed "accidents" between right- and left-leaning kissers. "I assume that the couples I observed did not kiss for the first time and were already adapted to each other," says Güntürkün. A direct kiss on the lips between couples facing each other is only possible, if both turn in the same direction.

An interesting question that remains unsolved is why the difference between right- and lefthanders, at 8:1, is much more pronounced. Güntürkün supposes that, apart from genetic causes, additional factors are responsible for this imbalance. "These may well be cultural," says Güntürkün. Many children are in fact educated to become right-handed.

DECISIONS IN THE BRAIN BECOME VISIBLE

Which part of the brain decides whether the left or the right hemisphere should do the job at hand? Or which area of the brain should address a given task?

The brain researcher Gereon Fink of the Jülich Research Centre and the University of Aachen, Germany, was the first to be able to observe this decision process. Together with colleagues at Jülich, Düsseldorf, London and Oxford he reported in Science magazine that a structure in the frontal area of the brain assigns each hemisphere its tasks. The researchers hope that their results will help patients who have suffered damage to one half of the brain, *e.g.* after a stroke.

The researchers asked the subjects of their experiment to look at short nouns, in which one letter was coloured red. Then, the participants were given different tasks. Sometimes, they had to state whether the word shown contained the letter A – a language task. On other occasions, they were asked whether the red letter appeared to the left or to the right of the centre of the word – this time, spatial perception was required. Meanwhile, the researchers observed which parts of the brain were particularly active. To this end, they used functional magnetic resonance imaging (fMRI). This method measures, how well the brain tissue is provided with oxygen; thus it makes those areas visible which are currently working intensively.

When they asked for the letter A, the solution of the problem was exclusively processed in the left hemisphere, and particularly in an area known as the Broca region, which has long been known to play an important role in language processing. However, if the location of the red letter had to be assessed, the same word would only trigger activity in the right half of the brain, and specifically in the apical lobe.

The researchers did not content themselves with just observing this division of labour, as they wanted to find out how the brain assigns the work to either of the two hemispheres. For this management task, the brain needs a control centre, which the researchers also located with the help of fMRI. They found that a certain area at the front of the brain, known as the anterior cingular cortex or ACC, decides whether the left or the right half of the brain becomes active. As Klaas Stephan from the Jülich Research Centre explains: "The left part of the ACC collaborated intensively with the language region of the left hemisphere, while the decision in favour of letter recognition was made. In the other cases, the influence of the right ACC on the apical lobes of the right hemisphere increased.

With this work, the researchers could monitor for the first time how the different regions of the brain communicate with each other while they assess a problem and find out who should be in charge. "We can see how the different regions that are involved talk to each other, and how the 'conversation' changes when the task changes," explains Gereon Fink.

Such insights can also help us to understand what happens in the brains of people who have lost this control mechanism, for example after a stroke. Thus, damage in the right apical lobe can make patients ignore one half of the world around them. They see that part, but don't pay any attention to it – researchers call the phenomenon "neglect."

Some patients with strokes in the left hemisphere, in contrast, can no longer understand language properly – a disorder known as "aphasia." In both cases the communication between different areas of the brain is inhibited. The researchers can now follow these "management problems" of the brain in detail – a prerequisite for eventually being able to develop better therapies.

3.4 Can Thoughts be Measured?

Couples like Bianca and Michael sometimes appear to have the same spontaneous thoughts. "You took the words out of my mouth," one of them might say, while both look at each other in amazement. It probably happens to most of us. Are these simple coincidences favoured by obvious conclusions? Or are there mysterious "psi phenomena" at work, which occasionally achieve a coordination of thoughts? We would prefer not to speculate any further, lest we find our book on the esoteric shelves of the bookshops.

Instead of inter-brain communications, we would like to investigate the burning question whether and to what extent brain processes can be detected and measured with the methods of modern science. Before we can do that, however, we need to have another, more comprehensive look at the neuronal architecture of the brain.

In many respects, the nerve cells of the brain are just like the other cells of the human body. For example, each brain cell has a nucleus hosting its DNA, the cellular power plants known as mitochondria, and a membrane that wraps up the entire cell, along with other cellular substructures. The most important difference between nerve cells and most ordinary body cells is that the former do not divide any more after the end of the embryonal development. In other words, the supply of nerve cells available at birth will have to last for a lifetime!

The second major difference are the long fibres that protrude from the body of a nerve cell and enable it to communicate with other cells. Only one of these fibres, the axon (from Greek word for axle) is able to convey information to other cells. All the other fibres emanating from a nerve cell are dendrites (from the Greek for tree). They have the task to receive information from the axons of other nerve cells. While the axon can grow to a respectable length of up to a metre, dendrites tend to be very short. They rarely reach a length of one millimetre.

Axons communicate their impulses by secreting chemical messengers into the synapses, the narrow gaps that separate them from the dendrites, which on average belong to 1000 (up to 6000) different neurons. These neurons are connected to one another in local circuits within the different regions of the cortex, thus forming the lowest level of neuronal architecture. Depending on the tasks they have to fulfil, very different connection patterns will be established.

3.5   A Formula 1 Racing Car in Michael's Brain

Which is the force that triggers the impulses? What happens in the brain cells in physical-chemical terms? In order to answer these questions, we will have an even closer look at the axon. In a simplified representation, we can consider the axon as a long, thin tube covered by the membrane of the nerve cell. Within the membrane, we find the internal substance of the axon, while outside there is only the extracellular liquid of the tissue.

"The inner substance and the external liquid have very different chemical compositions," explains the US biologist Robert Ornstein, who is the president of the Institute for the Study of Human Knowledge at Los Altos, California. The substance inside normally contains many protein molecules and very little sodium. Ornstein says that the most important event in the nerve impulse is the flow of sodium ions through the cell wall, from the outside to the inside. The ions travel through ion channels, which are like tubes crossing the cell membrane. These channels are usually closed, but when a nerve impulse arises, they open up for a short time and allow the sodium ions to enter the cell.

The speed with which electrical impulses travel around the brain is difficult to imagine. Typically, they race with the top speed of a formula 1 racing car – some 200 miles per hour – from one cell to the next. In order to make this possible, many thousands of ions have to pass through each of the channels in a thousandth of a second. Even for you to turn a page of this book, around two billion ion channels in your brain have to open and close again, just to carry out this very simple movement of the arm muscles. Considering that Michael put much more effort into hugging Bianca at the airport, the number of ion channels activated in the process must have been even higher.

4 THE CELL AS A CHEMICAL REACTOR

Technically speaking, each cell can be seen as a chemical reactor. The wall of this reactor consists of the above-mentioned cell membrane, which separates the interior of the cell from the environment. This way, the cell can create a variety of conditions for an equally large variety of biochemical reactions, and it can ensure that these reactions smoothly run their course.

However, cells are a bit more complicated than this simple image suggests, as there are also membrane-enclosed areas within the cell. These have quite important functions, such as the generation of energy or the synthesis and degradation of membrane proteins. Materials and signals must be transported across these membranes; these are important tasks carried out by specialised proteins embedded in the membrane. Similarly, the above-mentioned ion channels are specialised proteins in charge of the specific transport of ions, such as sodium or chloride ions.

Ion channels are often grouped according to which types of ions they let pass, such that one distinguishes between potassium, sodium and chloride channels, for example. By opening water-accessible pores within the membrane, they allow passive transport to proceed by diffusion, *i.e.* by the natural movement of all molecules. As ions carry electric charges and pass the channels quite quickly, this will result in a considerable electrical current. Such currents are the basis of the electrical data processing in the nervous system and the muscles. The examples of electric eels and those species of ray that can use their electricity to stun other fish show how strong these currents can be. Thomas Jentsch of the Centre for Molecular Neurobiology at the University of Hamburg has shown in his research that these animals contain large numbers of chloride channels in their electrical organs.

However, the substance transport across the ion channels is equally important. It is vital to establish the correct concentration of ions in certain parts of the cell; ion transport through the membranes along the surfaces of body tissues, the epithelia, are responsible for the transport of salt and water towards our vital organs, including the digestive tract, liver, kidney, and of course the brain.

4.1 The First Encounter With Bianca: Michael's Ion Channels "Remember"

Ion channels also play an important role in the registration of memories. Following a stimulus, they can undergo a lasting change and thus fix the information. Researchers have drawn this conclusion from experiments using the large sea slug *Aplysia*, which weighs between five and ten pounds and whose central nervous system contains only around 20,000 nerve cells. If a painful stimulus, such as a jet of water, hits the slug's head, it immediately withdraws its gills, trying to protect itself from the suspected danger. After around ten such stimuli, this reflex will be inactivated for about one hour. Thus, *Aplysia* has become used to the stimulus and stored the information in its memory.

In human beings, of course, there is a much more complex interplay of memory and the ion channels that take part in these processes. By now we know that apparently a precise sequence of molecular changes has to occur in nerve cells for us to remember an event, a melody, a smell, taste or gentle touch. This unique ability of our brain enables us to live "consciously", to carry out actions based on experience, and to mature (hopefully) during the course of our lives.

At the same time, these memories are also linked to emotions. For example, when Michael recalls his first encounter with Bianca, his feelings from that time come alive as though in a time travel. A certain melody – no matter whether classical or pop music – can also serve as a shortcut and trigger memories and emotions.

4.2 Michael has Localised Bianca: The "Chemistry of the Moment"

Without exaggeration one can say that ion channels are the cell's gates to the world. Controlled by chemical or electrical signals, they open or close like sluices, allowing external signals to affect sensory cells and starting cascades of reactions. This is also true for Michael's eyes, which immediately localised Bianca in the crowds. Therefore, we shall have a closer look at the chemistry of the visual process.

Like every healthy human eye, Michael's peepers contain different kinds of visual cells. Three types of cones react with particular sensitivity to specific parts of the optical spectrum and thus allow him to see colours. However, this process is only suitable for seeing things under satisfactory illumination, *e.g.* in daylight. At twilight, the 100-fold more sensitive rods take over, but they can only deliver a black-and-white image. That's why we appear to see things in greyscales at night (Figure 5).

In the dark, positively charged sodium ions enter the visual cells via its ion channels – this is known as the dark current. Under illumination, however, these gates close immediately. Therefore, the distribution between the interior of the cell and the outside changes, which in turn triggers an electrical signal.

Benjamin Kaupp, director of the institute for biological information processing (IBI) at the Jülich Research Centre, describes the function of the rods thus: "They are so sensitive that a single light quantum can be sufficient to close their ion channels. For this purpose, the light signal is amplified more than 10,000-fold by a cascade of biochemical reactions that requires a high energy input. Initially, the light triggers a re-arrangement of the visual pigment rhodopsin. The altered version of

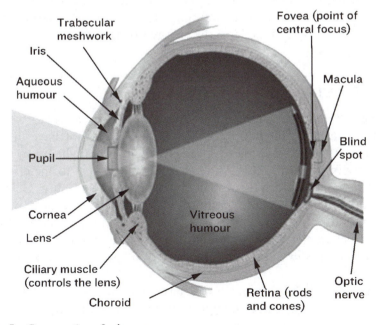

Figure 5 *Cross section of a human eye*

rhodopsin can then activate many copies of the molecule known as transducin, which in turn make an enzyme (phosphodiesterase) degrade a specific chemical messenger, the cyclic guanosine monophosphate (cGMP). As soon as this messenger is no longer available in sufficient quantities, the sodium channels close. The dark current is interrupted, and the cell announces the detection of light."

With his coworkers, Kaupp has searched for an answer to the question in which way Nature avoids squandering energy in the rods when there is sufficient light for cone (colour) vision. In their search they hit upon protein molecules which contain the building block glutamic acid in high concentration. Glutamic acid is (normally in lower concentration) one of the 20 amino acids that make up most proteins, and it also plays an important role in metabolism.

This protein molecule, known as "glutamic acid rich protein" or GARP, appears to play a prominent role in the chemistry of the visual process. As Kaupp found out, the mysterious GARP sticks like a leech to phosphodiesterase and thereby inhibits its activity. Furthermore it links up with another enzyme, which produces the messenger cGMP.

"Thereby a circle is set in motion," explains Kaupp. As a "doorman" or "messenger" the GARP molecule commutes between the outer cell membrane and the inner membrane stacks, which harbour the rhodopsin. Kaupp thinks he has obtained an important insight into chemical aspects that were previously unknown. His hypothesis: "Presumably, GARP holds back all the participants during daylight, until the rods are needed as dusk falls. When it is dark, the energy-saving complex falls apart."

A LIE DETECTOR FOR LOVERS?

When his science fiction opus "2001 – a Space Odyssey" was premiered in 1968, director Stanley Kubrick could not have dreamed that parts of this expensive spectacle might be caught up by reality exactly in the year 2001. And yet it happened, due to the efforts of two researchers from the technical university in Hamburg, Germany, Dietmar Schröder and Bernhard Fuchs. They were not so much interested in the secrets of Jupiter, but in the interiors of the movie's "main character," the computer "Hal," which maintained a physical and mental connection with the crew of a spaceship on its travel to Jupiter.

Although Schröder and Fuchs have not yet presented a Hal prototype, they have made an important step in that direction. "We have succeeded in making significant progress in the processing of bioelectrical signals," as Schröder describes the key innovation. While so-called biocontrollers, which accept bioelectrical signals as input instead of mouse, keyboard or joystick, are already on the market, they are usually limited to the relatively simple recognition of eye movements of muscle contractions.

The German researchers have developed a new and much more sensitive kind of communication. With a technical trick, they can now process typical bioelectrical signals like brain waves and heart rhythms much more efficiently and sensitively than ever before. The invention also represents a milestone for telemedicine, whose bottleneck had so far consisted in the very limited resources of the mobile measuring instruments, which were no match for the power and capacity of the stationary part. Therefore, the Hamburg researchers envision instruments for the recording and transmission of bioelectrical signals, such as electrocardiograms in the shape of miniaturised, portable units, as the first application of their invention.

"But the method is also suitable for processing a multitude of other bioelectrical signals," explains Schröder. These include for example the extremely weak signals which arise in the body in connection with metabolic processes. Similarly, the states of being in love or pangs of love should be detectable via the secretion of specific hormones. Nevertheless, Schröder and Fuchs are not planning to equip registry offices with the instrument in order to test brides and grooms before they say "I do" in the manner of a lie detector. Infidelity also remains beyond the scope of their scientific investigations.

Fortunately, their system – unlike Hal – cannot recognise thoughts and feelings in detail. However, the immediate prospects for the future are no less spectacular. With the help of microscopically small sensors, which record characteristic data such as body temperature, pulse frequency, blood pressure, *etc.*, the health status of a patient can be assessed online. It is imaginable that in the near future, certain groups of patients will be supervised in their own homes, as the measured data are transmitted first to their GP and from there possibly to specialised centres for remote diagnosis.

4.3 Michael's and Bianca's Emotions are Mainly Happening at the Front of the Right Hemisphere

In the view of the Austrian psychologist Werner Stangl, such specific functional systems exist for all sensory organs (as they are required for the interpretation of sensory input), and for organs whose movements are controlled by the brain. "This also applies for all vital functions that are controlled unconsciously" he adds. Furthermore, there are regions for the processing of language and concepts, for logical-rational thinking, and a completely separate system for decision-making. The latter is strongly correlated with the ability to deal with emotions, the processing of the body's sensory perception, and social behaviour, and it is usually located at the front of the right hemisphere, says Stangl.

Although it may sound incredible, it is already possible to visualise a crude representation of the information flow in the brain. "Today we know that brain cells practically 'fire' in a coordinated way," explains Peter Tass from the Medical Institute of the Jülich Research Centre. Together with his coworkers he has succeeded in recognising the most interesting patterns of neuronal activity among the countless signals originating in the brain.

He claims that the recognition of pictures depends on the synchronous neuronal activity of the different areas involved. "Synchonisation of neuronal activity is a fundamental mechanism of information processing," says Tass. Without short-term couplings within a group of nerve cells or between neurons of different areas of the brain, there would be no exchange of information, and thus no cognitive processes like learning or memory.

In other words: When Michael's brain creates his own individual world regardless of the hectic surroundings of the airport, the synchonicity of the neurons is an important prerequisite for this.

Synchronisation also plays a role in many disorders, as Tass points out. Thus, the uncontrollable trembling, known to medics as a "tremor," in patients with Parkinson's disease, is also a consequence of synchronous neuronal activity. Similarly, the disrupted movements from which patients with dystonia suffer fall into the same category, says Tass.

4.4 "Squids" Made of Niobium as a Compass of our Thoughts

The key to the success of Tass's group were so-called SQUIDs. Not the squids of the eight-legged kind, but Superconducting QUantum Interference Devices. This mouthful and its more memorable acronym stand for a novel sensory system, which is used for the detection of the extremely weak magnetic fields generated by small electrical currents. The motivation behind the development of such an extremely sensitive "compass" was the idea that these magnetic fields might offer a non-invasive way of monitoring the currents associated with the transport of information in the brain.

At the heart of the highly sensitive SQUID compass there is a superconducting alloy of the metal niobium, which is more commonly used as an anti-corrosion component in steel. And it is a cold heart indeed: liquid helium keeps the niobium at just four degrees above absolute zero. Thanks to the superconductivity at this low temperature, the SQUID can detect magnetic signals which have only a millionth of the strength of the Earth's magnetic field.

The resulting magnetoencephalogram (MEG) differs from the "conventional" electroencephalogram (EEG), which detects electical potentials at the surface of the skull, mainly in its improved spatial resolution. As the magnetic fields pass the skull without any change to their directionality, brain areas with rhythmic activity can be localized to within five millimetres.

In order exclude disturbances from heartbeat and eye movements, researchers typically record an electrocardiogram (ECG) and an electrooculogram (EOG). The latter is a method by which the voltage difference between the back and the front pole of the eye are used to record eye movements. An additional scan using magnetic resonance imaging (MRI) technology provides exact anatomical information. During this scan, the patient is placed in a strong magnetic field and exposed to high-frequency radiowaves. These waves can lift the hydrogen atoms in the brain to an "excited" state, which in turn can send out electromagnetic waves, which can be recorded with external sensors.

"Until now, the MEG method has not been able to establish itself in the clinical practice, as the measurements are quite involved and expensive," explains Tass. Apart from that, the interpretation of the data is also difficult.

The latter problem could be overcome, however. In order to be able to study neuronal synchronisation processes in ways that are gentle to the patient, using MEG, Tass has just developed a novel mathematical method for the statistical analysis of the measured data. This enables researchers to obtain information from human subjects which was so far only available from animals with implanted electrodes.

5   OUTBURSTS OF EMOTION SET MOLECULAR "SUBMARINES" GOING

In order to detect those events that are ultimately responsible for the chemistry of love, however, we must advance much deeper into the brain. With the pituitary gland, we reach the last and the decisive level of brain function. What we find there is a microcosm in which selected molecules glide through the blood stream like submarines, only to cast anchors at well-defined locations, where they develop their activity. It is a world governed by biochemistry, which ensures that billions of nerve cells can function without problems. This chemistry not only governs our vital functions, but also our very own thoughts and feelings.

The relationship between Bianca and Michael, too, is ultimately governed by such molecules. In fact, one could say that an intricate chemical system is in charge of the various feelings of love. Even the libido (the sex drive) is essentially controlled by chemistry.

Has science already intruded into this most intimate area of human life? In principle yes, but the situation is similar to a well sealed love letter. You may be able to identify sender and recipient, but from this you cannot draw conclusions to the actual content of the letter.

Similarly, the researchers at Jülich, in collaboration with others at the University of Düsseldorf, have succeeded in identifying those brain cells that are currently sending or receiving molecular messengers. With the help of modern brain imaging methods, the researchers can visualise the processes going on inside a person's head. Their signposts are the above-mentioned electrical impulses, which run along the nerve fibres and give rise to magnetic fields. Moreover, when the finely dispersed capillaries transport more blood to the active areas, this can be detected, because the increased consumption of the fuel glucose also uses up more oxygen.

For such investigations, the researchers at Jülich use functional magnetic resonance imaging (fMRI), which maps the oxygen content of the blood. Depending on how much oxygen the hemoglobin carries, there is a different type of "echo." Using this approach, researchers have identified areas of the brain which are involved in motion control and in decision-making (Figure 6).

The Jülich research group has developed and patented a new method known as FIRE (Functional Imaging in REal time) which speeds up

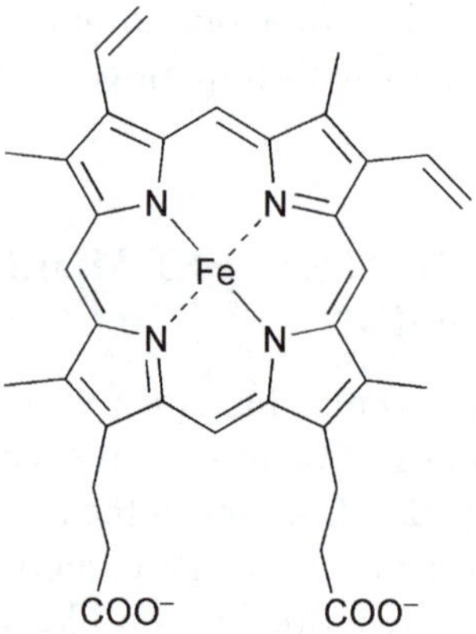

Figure 6   *The heme group of hemoglobin*

functional MRI to the extent that images can be generated as quickly as changes occur in the brain, i.e. in real time. One might have called it turbo MRI.

5.1 Michael's Love Triggers an Avalanche of Signals in the Brain

The researchers have already gone one step further. With the help of real time MRI, they can identify and quantify the molecules involved in brain activity. Thus they can distinguish whether a signal they measure arises from a messenger compound of the brain cells or from an energy transferring phosphorus compound.

If they could have analysed Michael's brain at the moment of Bianca's arrival by real time MRI, the researchers would certainly have spotted a whole cascade of signals. Fortunately, though, even then the content of the messages would have remained secret, as the identification of the messenger compounds allows them, figuratively speaking, to glimpse the shape and colour of the envelope, but not to read the message it contains.

These encoded messages are transferred by messengers like neurotransmitters and neuropeptides. How rapidly a given piece of information is passed on depends on which kind of nerve cells is involved, on the neurotransmitter, and the number of synapses (the above-mentioned connections between nerve cells) that take part in the process.

The various transmitters are present only in small amounts, and they are not distributed randomly in the brain. Rather, the nerve cells that work with the same transmitters, tend to be organised in local groups, and their fibres radiate only into well-defined areas of the brain. One example for this is provided by the nerve cells that work with the transmitters dopamine, serotonin, and noradrenaline. These transmitter compounds belong to the group of the monoamines, while others, such as gamma-aminobutyric acid (GABA) are amino acids. The most important neurotransmitters and their mode of action are described in the Box below.

THE MOST IMPORTANT NEUROTRANSMITTERS

- Acetyl choline (or (2-acetoxyethyl)-trimethyl ammonium chloride): This transmitter binds to the so-called nicotine receptor. For an instant shorter than a blink, just 0.1 milliseconds, this ion channel opens and allows 15,000 to 30,000 ions to pass. Acetylcholine has a stimulating effect on the synapse between nerve and muscle, because it allows the positively charged sodium ions to flood into the cell.

- Glutamic acid (2-amino glutaric acid): This amino acid acts on different receptors influencing the influx of sodium ions, the efflux of potassium ions, and in some cases the influx of calcium ions. As an "excitatory amino acid", it is one of the most important transmitters in the central nervous system.
- Serotonin (5-hydroxytryptamine): This neurotransmitter, which was first isolated from blood serum in 1948, plays an important role in sleep-wake patterns. Furthermore, it has a decisive influence on our moods. As an indole derivative, serotonin is structurally related to some hallucinogens, such as psilocybins (in "magic mushrooms") and LSD. Also see p. 82.
- GABA (gamma amino butyric acid) causes an influx of chloride ions and thereby an inhibition of the receiving cell.
- Glycine (amino acetic acid): The simplest neurotransmitter (and simplest amino acid) occurs throughout the body and acts similarly to GABA, inhibiting cells via an influx of chloride ions.
- Dopamine (4-(2-aminoethyl-)brenzcatechol), together with the functionally related:
- noradrenaline (4-(2-amino-1-hydroxy ethyl)-resorcine) acts indirectly by increasing or decreasing the concentration of a second messenger, which then in turn triggers the electrical or biochemical effects. The US biochemist, pharmacologist and physiologist Earl Wilbur Sutherland identified this second messenger, cyclic adenosine monophosphate (cAMP), for which he was awarded the Nobel prize in physiology or medicine in 1971 (Figure 7).

6   NEUROTRANSMITTERS – GUARDIANS OF EMOTIONS AND SLEEP

Nerve cells that use dopamine as a neurotransmitter are located mainly in the middle of the brain and reach through to the frontal brain, where they are involved in controlling emotional reactions. Other dopamine-containing nerve fibres end close to the centre of the brain in the striatum, where dopamine plays a role in the regulation of complex movements. Degeneration of these nerve fibres has devastating consequences and leads to muscle rigidity and trembling, among other problems.

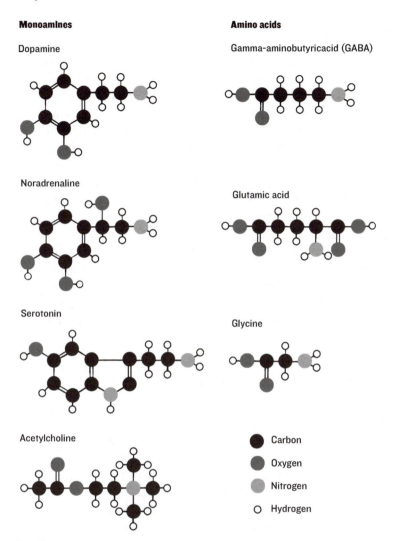

Figure 7 *The most important neurotransmitters*

Serotonin containing nerve cells are located in a small part of the brain stem, known as the raphe nucleus. The fibres of these nerve cells reach into the hypothalamus, the thalamus, and many other brain regions, where serotonin is involved in the perception of sensations and in bringing about sleep.

Noradrenaline containing nerve cells are found particularly in a small area (nucleus) of the brain stem called the locus coeruleus or LC. These nerve cells are highly branched and connected with the hypothalamus, the cerebellum, and the frontal brain, where the noradrenaline appears to contribute to the maintenance of wakefulness, the reward system of the brain, dreaming, and the regulation of mood. "Reward system" is what brain researchers call the centres for pleasure sensation. These play important roles in learning and also in memory functions.

Apart from the transmitter chemicals mentioned above, scientists know or suspect further compounds of carrying out similar functions, including some of the amino acids. For example, glutamic and aspartic acid have a stimulating effect on most nerve cells. The simplest amino acid, glycine, is known to act as an inhibitory transmitter in the spinal cord.

The most important inhibitory transmitter is gamma-amino butyric acid (GABA), which is synthesized exclusively in the brain and spinal cord. It has been estimated that a third of all brain synapses use GABA as a transmitter.

Small difference – big effect: It is remarkable that the inhibitory transmitter GABA differs from the stimulating transmitter glutamic acid only in the small detail that it lacks the second carbonic acid function, so it has just one carbon and two oxygens less than GABA. Thus, subtle differences in the molecular structure can make transmitters act in opposite ways.

6.1   Is Michael's Brain Squandering Energy?

Apart from the seven classical and in chemical terms relatively simple neurotransmitters, researchers have discovered around 50 additional chemical messengers which are now also known to act as neurotransmitters in the brain. Over the last few years, both the characterisation of the transmitter compounds and the analysis of the molecular processes have made remarkable progress.

"Thanks to these investigations, it is now known that the effects of many pharmaceuticals and neurotoxins are based on the fact that they interrupt, modify, or mimic the chemical transmission of nerve signals, and there are indications that mental disorders are also caused by defects in the function of chemical transmission systems in the brain," as Oxford pharmacologist Leslie Iverson summarises.

He also says that the brain is one of the biggest consumers of energy in our body, witness its strong blood circulation and high requirement of oxygen. Although the brain accounts for only two percent of the body weight of an adult, says Iverson, it benefits from 20% of the oxygen intake at rest.

Therefore, today's brain researchers suspect that the brain's high energy requirement results from the necessity to maintain the concentration gradient of ions at all nerve membranes. The generation, processing, and transport of nerve signals in our many billions of nerve cells depends on these gradients. For the brain there is no breather. In fact, the intensity of the brain's metabolism is roughly equal day and night. It may even increase slightly at night, when Bianca and Michael are dreaming.

Trying to investigate more closely the intensity of the working brain, the researchers hit a few limitations. Of course it is possible in principle to study neuroreceptors in the test tube. To this end, the scientists chop up the nerve cells isolated from living or dead brains, until only their coats, the membranes with the embedded receptor molecules remain. Then, the individual chemical building blocks from which the receptors are made can be identified. The only problem is that at that stage the function of individual receptors and their complex interaction with other types of receptors in the living brain can no longer be observed.

One solution to this apparent dilemma is offered by a method known as positron emission tomography. This technology makes it possible to visualise the different receptors in the brain with the help of radioactively labelled molecules. But how can one persuade the brain to ingest these substances?

6.2 Bianca Likes Glucose – Her Brain Too

By now, researchers have actually succeeded in making the intensity of the energy metabolism of individual brain cells visible. This required a trick, and the brain's appetite for glucose showed them the way.

Are you by any chance among those people who develop an increased appetite in stressful situations and tend to nibble at sweets? If so, you should not feel bad about it, as you are in good company, as 50% of the authors of this book display the same tendency.[†] Nerve cells, too, are literally getting greedy when they need more energy, and they consume more glucose than when they are at rest. The glucose they take up is normally turned over quite rapidly, such that after a short time it will no longer be detectable.

The situation presents itself in a different light, however, if the glucose is equipped with a tracer. For this purpose, researchers at Jülich have labelled glucose with the radioactive isotope of fluorine, ^{18}F. The distribution and use of glucose labelled in this way can be monitored from outside the body by positron emission tomography (PET).

[†] In fact, people tend to respond differently to exams or other stress situations. While the female half of this team of authors tends to forget the food supply, the male author visits the refrigerator with conspicuous frequency. Unfortunately, the bathroom scales don't honour mental gymnastics in the least, and a family trade-off has also failed to materialise.

HOW TO MAKE A RADIANT GLUCOSE LOOKALIKE

First, one has to produce radioactive fluorine in a particle accelerator. Water molecules containing the isotope ^{18}O are bombarded with hydrogen nuclei (protons). The resulting ^{18}F- fluoride ion runs through minute capillaries into a reaction chamber encapsulated in lead. There it reacts with glucose, forming the labelled molecule, which still carries several protective groups. After a total process time of less than an hour, the radioactive lookalike is ready to use.

For investigations of the living brain, labelled glucose is an invaluable help, even though it will most probably never be used to investigate the degree of being in love via the activity of the brain.

However, it is being used in order to assess the damage after a stroke. As the labelled molecule passes the blood-brain barrier just as well as ordinary glucose, it is a valuable indicator showing where the metabolism of the brain tissue proceeds normally and where it is hampered by a blocked blood vessel. The higher the glucose consumption of the affected brain region, the better are the chances of recovery.

By now, many hospitals receive the labelled glucose from Jülich, which in 1999 was officially licensed as the first radioactive diagnostic for PET examination in Germany. As the radioactive fluorine decays relatively rapidly – half the radioactivity is gone after 110 minutes – the radiation dosis the patients are exposed to corresponds to that of an ordinary X-ray examination.

Therefore, the product has recently also found a new use as a cardiac diagnostic. Thus, after a heart attack, doctors can find out which areas are irretrievably damaged and which can still survive.

Similarly, radioactive sugar is about to revolutionise cancer diagnostics. The labelled glucose is enriched preferentially in the tumours, as they maintain an accelerated metabolism in order to fuel their rapid growth. With the help of PET cameras, brain tumours can be located with millimetre precision.

CHAPTER 2

Two Networks of Nerves

Our subjective inner life is what really matters to us as human beings.
Benjamin Libet (neurophysiologist)

1 HUGS AND HEARTBEATS

Let us return to the point where Bianca and Michael ran towards each other, and hugged tightly. As we have seen, billions of messages have been transmitted via ion channels in Michael's brain, in order to bring about the conscious and intended movements of his muscles. But at the same time, his body reacted in a different way, over which his conscious self had no control.

"Your heart beats like crazy," Bianca noticed immediately. Tenderly she felt his skin under his T-shirt. "And you're sweating a bit, too."

"Does that surprise you?" replies Michael, breathing heavily. "No," says Bianca with a loving smile, "same here."

These very different actions of the body – conscious movements versus unconscious reactions like sweating, racing heartbeats, fast breathing, *etc.* – reflect the dualism of our nervous system. Nature has equipped the human body with two systems, known as the somatic and the vegetative nervous system.

The somatic nervous system is largely under the voluntary control of a person, *i.e.* we can guide it consciously. With its help we coordinate our movements, such as lifting a hand, turning the neck, and complex activities like walking or swimming. The muscles used in these activities are called the striped musculature.

1.1 Control Over Sex and Senses

In contrast, the vegetative nervous system is also called the autonomic nervous system, as its effects are not controlled by our will. It regulates our vital functions including breathing, digestion, metabolism, secretion, and water management. Furthermore it also controls organs and organ systems such as the sexual organs and the inner eye musculature.

For a more precise analysis of these functions, it is necessary to further subdivide the vegetative nervous system, namely into the sympathetic and the parasympathetic system. This has nothing to do with sympathy. The sympathetic nervous system is defined by the fact that its nerve endings release the neurotransmitter noradrenaline, which triggers the body's adaptation to a higher level of performance or activity. Thus, the influence of the sympathetic system makes the heart go faster, the channels in the lungs grow wider, and any bowel movements are inhibited.

In contrast, excitation of the parasympathetic nervous system, which uses acetyl choline as a neurotransmitter, leads to a reduced heart frequency, to a narrowing of the breathing channels, and an increase of activity of stomach and intestines, rather corresponding to a state of rest and digestion. Consequently, humans tend to have a higher sympathetic activity during the day, while parasympathetic activity dominates the night.

Functionally, the two systems can be described as antagonists, which means that their influence on a specific organ is normally of opposite nature. Nevertheless, the vegetative nervous system often cooperates with the somatic system in order to maintain the important balance of the variouis organ fuchtions. For example, after a drop in temperature, the hypothalamus will order the vegetative system to make the blood vessels close to the body surface narrower, to minimise the loss of heat, and at the same time send signals to the somatic system, to trigger muscle tremor.

1.2 Team Work

In order to better understand the different functionalities of the two systems, let us consider the following situation: A man is exposed to immediate danger and trying to escape. Therefore, he needs more blood in the muscles, and thus a higher pumping frequency of the heart.

His sympathetic system is working at its highest rate. He begins to sweat, as the sweat glands are activated by the sympathetic system. His lungs will work harder to get more oxygen into the blood. When the danger has passed, he will settle down, his heart rate will decrease, the sweating will cease, and the lungs will take in less oxygen.

All these instructions come from the parasympathetic system, which returns the body to a relaxed state. This example has shown that the sympathetic system puts our body and psyche in a general state of alert, while the parasympathetic system winds down the alert functions and brings about a restful state.

Even though we cannot consciously control the vegetative nervous system, autogenous training allows us to influence it within certain limits. It is difficult to define exactly the distinction between what can or cannot be controlled. Although it is generally true that the vegetative nervous system is beyond our control and consciousness, we can in fact notice the symptoms of its intervention. Moreover, the vegetative system can be influenced very strongly by psychological factors. This is why autogenous training enables us to achieve at least partial control over its functions.

1.3 No Erection without the Cooperation of Vegetative and Somatic System

Despite their differences, vegetative and somatic nervous system work closely together. Especially in the upper brain stem, the hypothalamus, and the cerebrum, they can no longer be separated unequivocally. At the periphery, *i.e.* outside of the CNS, these two systems can normally be clearly distinguished both anatomically and by their effects.

One example of the cooperation of the vegetative and somatic system is the erection of the human male, which can be induced either unintentionally (*i.e.* by external stimuli beyond the person's control) or by conscious thought processes.

And while we are touching this subject, we would also like to answer another important question: What happens during an erection?

In fact, the erection is based on a very complex process, in which both the nervous system (brain and spinal cord) and the blood vessels of the penis are involved. Of particular importance are the blood vessels of the erectile tissue, which are especially numerous in the sponge-like tissue of the penis. In the flaccid state, the incoming vessels are constricted, such that only little blood flows into the penis.

When a man is sexually stimulated, nerve signals make the blood vessels in the penis widen. Thus the blood flow to the penis increases. At the same time, the erectile tissues, as they fill up and swell, compress the veins that carry the blood away from the penis. This is an important process, which is crucial for maintaining the erection (see Chapter 18 for details on how drugs can help with this). By the way, the physiological details concerning the stimulation of a woman's clitoris are very similar.

Obviously, the erotic stimulation is also connected to a range of physiological changes, as the vegetative nervous system responds immediately to the stimulus. Higher pulse rate, higher blood pressure, and a slightly increased body temperature are the corresponding symptoms. At the same time the "patient" undergoes a partial loss of his (or her) perceptive abilities.

CHAPTER 3

Signals of Love

"Love is when the desire to be desired attacks you so violently that you
believe you will die from it."
Henri de Toulouse-Lautrec (Painter, 1864–1901)

1 LOVE AT FIRST SIGHT?

"When I first saw her, lightning struck" – one of many descriptions of
the phenomenon known as "love as first sight." For Bianca and Michael
it was similar, when they first met at a party of a mutual friend. And
today they also know that it was more than just a flash in the pan.

Our senses are the gateway for the stimuli that we are exposed to, be
they visual or acoustic, taste, touch, or smell. All these kinds of information are selected, transmitted and processed in a complex interplay
involving electrical fields, neurotransmitters and hormones (which we
will cover in detail in the following chapters), and eventually perceived
and fitted into the world that is our personal experience. And when love
at first sight strikes a person like lightning, it is obvious that their
vegetative system is highly involved. In other words: any resistance, if at
all desired, may become difficult.

At least science has already started serious investigations of love at
first sight. For example, the research institute for human ethology of the
Max-Planck Society at Andechs, Germany, has revealed some details.
Apparently, love at first sight exists if the sighting lasts for more than 30
seconds, and women are much more likely to risk this prolonged eye
contact than men.

The researchers Karl Grammer and Christiane Doermer-Tramitz
gained these insights with the help of 300 pupils from secondary schools
in and around Munich, who were around 18 years old and played along
in a so-called flirt experiment. During a seminar, one of the researchers
asked a boy and a girl, who didn't know each other, to go to another
room, on the pretext of analysing a video recording. Then, however, the

researcher left them alone to answer an urgent phone call. The pair stayed in the room for 10 minutes, while being filmed with a hidden camera.

Afterwards, each of the pair had to own up to their feelings with the help of a questionnaire. How attractive they considered the other person, how confident and strong (or otherwise) they considered themselves, how strong the mutual interest in each other was, and (if there was an interest) how big the fear of being rejected. These replies were then compared with the recordings of the hidden camera, with the researchers concentrating mainly on the language and eyes of the pairs. From this analysis they found that the 30 seconds of the first eye contact normally decide the further development of the encounter.

The results showed among other things that the more attractive a woman rates a man, the sooner she will make eye contact. She talks more, asking questions of a general nature. However, the fact that she likes the man also makes her insecure, which becomes obvious from the way she speaks, using fillers such as "err," and committing grammatical errors or leaving sentences unfinished.

If a man takes a shine to a woman, he will display similar behaviour. He, too will make eye contact and signal his interest by asking lots of questions, which also reinforces her interest in him. The less a man talks, the investigation suggested, the less attractive he appears to the female partner in the test. However, his initial silence does not necessarily mean that he doesn't like the woman. Men only become talkative when they are confident. If he lacks confidence, it can happen that he stays silent even with a woman he is interested in, or that he only manages a stumbling conversation, in which he appears withdrawn and also avoids eye contact.

1.1 Shy Men Avoid Eye Contact—Women Look for it

The German researchers also found out that low confidence expresses itself differently in women. They, too, have trouble getting a conversation flowing with the man they fancy. Their sentences may be short and incomplete. However, even shy women try to establish eye contact with him, in order to demonstrate their affection. In other words, women in such situations are not as reticent as men.

Finally, whether somebody takes a fancy to another person, can also be decoded from the way in which a man or a woman start a conversation: with a question if they are interested, or with a meaningless comment ("It's cold in here!"), if they are not. Mutual sympathy, as a step that goes beyond interest, reveals itself through smiles, longer eye contact, and short expressions of affirmation or enthusiasm – the precise nature of which may depend on the age group a person belongs to.

2   ANATOMY OF A FLIRT

The psychologists Dirk Blothner and Barbara Contzen at the University of Cologne, Germany, wanted to know what's behind the flirting game, the playful exchange of erotic innuendo. They questioned around 50 participants in in-depth interviews to find out what the flirtatiously inclined get out of such games, and how they can be handled successfully.

Like other, more direct kinds of flirt, this game begins with an unconcealed erotic interest and may be accompanied by physical titillation and sensual phantasies. Long, deep looks with just a hint of risqué make suspense and longing come up between the flirting partners: the prickling sensation starts here! While they talk about the weather, the strong coffee, the unfriendly landlord, traffic congestion or work overload, their imagination knows no boundaries. However, there is also some skepticism in these desires. Perhaps it will go wrong; perhaps it will not play out as anticipated.

"This creates a back and forth, feeling your way, a play with double entendres," explains Barbara Contzen. In other words, the things that have an erotic effect, can on the other hand also be interpreted more soberly. For example, a man might in the course of a conversation casually put his hand on the woman's arm, which could be interpreted as a spontaneous gesture related to his storytelling. A similar function can be assigned to other, seemingly random, fleeting touches: the way of lighting the other person's cigarette, a gentle brushing over the shoulder, or the handshake prolonged by just a few seconds.

"In the flirt game we create space for our erotic desires, without having to take undesirable consequences into account," as Contzen explains this playing with the flames, designed not to cause a major fire. The double entendre serves as a protection against this outcome. One has to restrain any all too obvious pressure on either side, but also avoid cooling, as an end of the double entendres would also be the end of the game.

Blothner calls this the testing nature of the flirt game, where important gears of the psychological reality can be set in motion again, such as the male/female dichotomy, where "male" could for example serve as a shorthand for the more active and dominant side of life, and "female" for a more passive and conciliatory side. "Essentially, the question of 'male' and 'female' in this sense is renewed for every person in every moment," clarifies Blothner. In long-term relationships, however, these gears might become "frozen" in a certain position. With the flirt game, people can prove to themselves that they still know how to switch gears.

In other words, people who see themselves as more passive in their relationship can show more active characteristics in the flirt. Those who tend to be demanding, may enjoy being challenged for once. In this game, things are repeatedly turned upside-down. Sometimes, one side gains the upper hand, sometimes the other – a civilised form of the battle of the sexes in a protected framework, intended not to lead to any solidification.

However, the balancing trick isn't all that simple. "The problem is that an extraordinary intimacy is established, but at the same time a high degree of separation is maintained," explains Blothner. Paradoxically, people who live in a happy relationship are often most accomplished in this balancing trick. They can disengage themselves from their permanent partner for a limited time without fear, and they can also allow themselves to get carried away with another person, without having to fear that they might not be able to detach themselves again.

According to Barbara Contzen, flirt games succeed especially when they are based on a "dominant framework," *i.e.* an active environment that dominates the scenario. Such environments include train journeys, evening courses, the gym, or the laundrette. In the theatre, too, the framework conditions are so rigid that there is little risk of the flirt turning into something more serious, so the game can be extended. At the same time, the environment offers new material that can appear in a sensual or erotic light. Part of the attraction for the flirting couple is to see how even boring everyday procedures can gain sensual dimensions.

"The person that benefits most from the game is often the one who doesn't expect too much," says Blothner. People who are desperate to find a new partner often find flirting more difficult. Those, who do not want to endanger an existing relationship or who are happy to live alone, often come across as more relaxed, humourous, open and attractive. Because the flirt game, according to Blothner, includes the playful handling of engaging and disengaging behaviour.

3 THE AMERICAN KAMA SUTRA: 103 PULLING TACTICS

David Buss, a psychologist at the university of Ann Arbor (Michigan, US), put all prudishness aside and set out to explore how his fellow human beings are literally trapped by sexual stimuli. He questioned male college students about the most efficient method with which a woman might pull a man.

Asked about the effects of 103 different pulling tactics, the male students emphasized the more suggestive kinds of female behaviour, including rubbing chests and hips, seductive glances, pouting lips, blown kisses, putting her arms around a man's neck or stroking his hair, sucking a straw or a finger, standing very straight to emphasize her curves, or bending over to allow insights into her cleavage.

Similar behaviour from men, however, does not have any much effect on women, as Buss explains in his book "The evolution of desire: Strategies of human mating." The more unmistakable the sexual approaches of a man, the less attractive he will appear to women.

Using a 7-point scale, the psychologist tried to measure the efficiency of the approach. A value of one represented an hopeless attempt at chatting up, while a seven was given for the most efficient method.

For example, if a woman in a bar rubs her chest or hips against a man, how does that go down with the targeted person? A majority of the male participants in the study appear to appreciate this kind of approach, as it achieved an average of 6.07 points. On the other hand, a man should avoid approaching a woman in a similarly suggestive manner, as female students rated it only with an average of 1.82 points. Thus, explicit behaviour that has a highly stimulating effect on men (or at least on male college students who take part in such surveys), is seen as inefficient, if not repulsive by most women.

Another very efficient tactic for a woman in search of adventures is to emphasize her sensuality with her clothing and behaviour, as Buss points out. The male participants in his studies reported that the following tactis attract them most:

- tight-fitting, provocative clothing, revealing the body contours;
- a deep cut back or daring décolleté
- a strap that slides off the shoulder;
- a very short skirt;
- swinging hips
- provocative dancing;
- proud, long legged gait.

Thus, a woman using these sensual signals can be almost certain of male attention.

CHAPTER 4

Tied Up with a Double Helix

We used to think our future was in the stars.
Now we know it's in our genes.
James D. Watson (molecular biologist)

1 ALL IN OUR GENES?

"Tell me," Bianca says looking into Michael's eyes and running her right hand through his hair, "have you been thinking of me in all those weeks?"

"Oh yes," Michael replies. "Each and every day my thoughts were with you several times. How about you, you must have found many things to distract you, haven't you?"

"Yes, that's true. Charlene and Steve have done their best to keep boredom at bay," Bianca admits. "But you didn't miss out, you were always my invisible companion." She snuggles up to him. "By the way, the two of them send their regards, and they would like to meet you soon."

"Nothing wrong with that. Err – remind me in which way you are related to Charlene? I'd better start getting used to my new family, shouldn't I?" Michael looks at her, smiling.

"That's a long story. The connection runs via our great-grandparents. Charlene and I are second cousins."

Michael wants to know more: "Any similarities?"

"Not so much on the surface," Bianca replies. "But it's crazy how many things we have in common."

"Really?" "Yes. Charlene thinks it must be in the common genes."

"Your genes account for your wonderful brown eyes and black hair." Bianca returns the compliment: "I like your dark curls, too." She strokes his hair again.

"Think your genes get on with mine?" Bianca puts on her sweet pout. "Do you have any doubt about that?" Michael takes her hand and presses it gently in his. "No," she says softly.

"Talking about genes, I have taped an interesting TV program about the human genome for you ... " Bianca's kiss puts an abrupt end to Michael's sentence.

2 FROM PRIMORDIAL SOUP TO DNA

The TV program that Michael just mentioned must have been about the recent progress in genome research. When we talk about the chemistry of love and emotions, we must not forget our genes. For they, too, contribute to what we call our personality. Genes are much more than just a blueprint for an individual. Among other things, they also influence our sensory world. Whether it's vision, hearing, smell, taste, or touch – without the right genes, we would not be able to perceive the world around us, let alone fall in love.

Let us therefore leave our lovebirds to themselves for a while and direct our attention towards those mysterious molecules which we have inherited from our ancestors. How far do we have to go back to find the origin of genes?

Firstly, modern genome research has established that all humans – no matter which race – are identical in 99.99% of their genetic material. If we include our closest animal relatives, the chimps, the identity is still at nearly 99%.[†] And even with completely unrelated animal species, such as the fruit fly, we still have 60% of our genome in common. Researchers conclude from this that all life on Earth goes back to a common origin.

Let us go on a time travel and investigate the environment where life originated:

Four and a half billion years ago, the area of space now known as the Solar System looked very different from today. There was no sunlight, and the Earth, the Moon, and the other planets were missing too. Instead, there was a huge, lens-shaped cloud of gass and dust hovering

---

[†] The fact that we seem to resemble the apes so closely should not bother us too much. While the recently published genome sequence of the chimpanzee has confirmed that the individual "letters" of our DNA are exchanged in only one percent of all positions, the genome size of 2.7 billion letters suggests that there are around 27 million "small differences" at the base level. On top of that, there is a smaller number of large scale insertions, deletions, and rearrangements of genetic material. Most importantly, however, small changes in key positions can have huge knock-on effects, especially if the proteins affected are involved in regulatory tasks. For instance, accidental exchange of a single letter can lead to diseases like cancer or cystic fibrosis. Similarly, a small number of mutations can direct the embryonal development into very different directions, accounting for the large differences in the outward appearance, abilities and behaviour. Thus, 99% genetic similarity with the chimp is no reason for us to go bananas!

in the empty space. This cloud was more dense in its centre than on the outskirts, and continued to collapse into the centre under its own gravitational pull, while the centre emitted just invisible heat waves, but no light. This collapse must have gone on for millenia, until suddenly, visible light emerged from the centre. A new-born star shone out of the cloud: our Sun.

Behind the scenes of this cosmic theatre, we see that elemental hydrogen, the simplest and lightest of all chemical elements, was the protagonist of this show. At the centre of the collapsing cloud, pressure and temperature had risen to the point where the fusion of atomic nuclei becomes possible. This process still fuels our Sun, and without it, our planet would have no light, no warmth, and therefore, no life. This was the first act of creation, the emergence of a world governed by physics. It was the era of physical evolution.

As the cosmic drama unfolded, the shroud of mist began to clear. Shreds of it fell into the young Sun, as if to fuel the new fire. Other parts of the cloud defied the pull of gravity by circling the Sun in elliptical orbits. This part of the primordial matter originally travelled on a multitude of different, intersecting orbits. Model studies have shown that frequent collisions between theses particles created a kind of friction that forced them into orbits that are more or less in one plane, as the planets and most asteroids still are today. Swarms of small particles began to coalesce into planetoids, which eventually grew into today's planets. As the heat of the accretion process dispersed, the rocky planets like our own started to form a solid crust including the first continents on Earth.

Some 700 million years afer the Sun sent its first ray of light into the Cosmos, conditions here on Earth were still quite inhospitable. Oceans smelling of sulphur and ammonia lapped against bizarre-looking continents made of volcanic basalts. Wind, weather, and frequent volcanic eruptions kept remodelling the surface. Neither the chirping of a cricket, nor the cry of a bird travelled the airwaves – the Earth was still lifeless. It was the second act of creation, the era of chemical evolution.

Evidence for this process of chemical evolution based on simple compounds came from an experiment conducted by the chemist Stanley L. Miller – then only 23 years old – at the University of Chicago in 1953. He simulated the environmental conditions of primeval Earth in a laboratory setting. To this end, he mixed together a "primeval atmosphere" consisting of the gases methane, ammonia, and hydrogen, put it in a closed vessel together with some water (the primeval ocean), and provided energy by discharges from electrodes (simulated lightning). After a few days, he could detect a number of organic compounds such as fatty acids, carbohydrates, and most importantly several kinds of amino acids. These are the building blocks of today's proteins, which include the enzymes that are crucial for controlling the biochemical reactions in all living beings.

Later, similar experiments also yielded purines and pyrimidines, which are important components of the nucleic acids desoxyribonucleic acid and ribonucleic acid (DNA and RNA), which serve as information carriers in all organisms. With their ability to be replicated in identical copies, they provide the foundation for genetic inheritance. Thus it appears probable that these or similar compounds also arose in the early days of the Earth, providing a chemical foundation for the origin of life.

Follow-up investigations failed to reveal a clear mechanism for the origin of life, but suggested that inorganic minerals contained in the oceans may have played an important role. Some researchers believe that clay minerals rich in zinc were important catalysts, which specifically produced the kinds of amino acids that are used today. If this holds true, the oceans represented the primordial soup of life. From this soup, condensation reactions must have – in ways as yet unknown – produced more and more complex molecules and finally self-reproducing units, on which the principles of evolution (mutation and selection) could begin to act. This was the third act of creation – the era of biological evolution.

Three billion years ago, there were already living organisms. Evidence for this is found in the shape of fossil structures known as stromatolites, which point to the existence of organisms resembling today's bacteria. While there are claims for even older traces of life, the precise dating and interpretation of these finds is increasingly controversial as one goes back in time, and the precise time when life originated – maybe 3.7 to 3.8 billion years ago? – remains open to speculation.

Although the precise process by which life originated remains unclear, most theories agree in that it must have produced molecules which can be copied and which can in a very general sense be called genes. Such molecules are a prerequisite of life, as they make it possible to pass on traits to a new generation, and thus allow organisms to reproduce.

3   CHEMISTRY: THE SEED OF LIFE, MIND, AND EMOTIONS

Hence we must not be surprised that everything in Nature – living or lifeless – is made of the same chemical elements. Even when we use space probes and telescopes to explore the depths of the cosmos, we meet similar kinds of chemistry all over again. Be it the red desert areas on Mars, owing their colour to iron oxides, or the fantastically colourful clouds of Jupiter bearing organic molecules that remain to be investigated in detail – chemistry is everywhere.

A few years ago, when the space-based Hubble telescope discovered an inferno of gas and dust in the Orion Nebula, resembling the state of our own solar system some 4.5 billion years ago, humankind not only peered into the depths of the Universe, but also – for the first time – at the beginning of its own history. Some 700 or 800 million years later, when the chaos in the Orion Nebula allows the first planets to form, chemistry may – if the conditions are favourable – once more produce similar molecules like those that arose on young Earth.

Thus, it is obvious that humans never invented chemistry, as it is an essential part of Nature and by many orders of magnitude older than our kind. The importance of chemistry not only reveals itself in the fact that it is ubiquitous today. It has a much more fundamental dimension: chemistry was and remains the seed of life.

4   GENES ARE LIKE MEN

Our genes, too, owe their existence to the developments outlined above. They haven't come into this world fully formed. Through the course of evolution they became what they are today.

The German biochemist Ernst Peter Fischer described them provocatively thus: "Genes are like men. They laze about in the cell and wait to be served. On their own, they are absolutely helpless, and cannot even move without assistance, but in the end all life functions of a cell depend on the instructions that they give. The genes leave the work to the molecular women, known to biologists as proteins. The proteins not only care for their molecular men, the genes, but they also fulfil many other tasks and control all the reactions that are essential for life." Fischer concludes: "A gene is not a molecule that the cell owns. It is a molecule that makes the cell what it is."

But how can we imagine a molecule carrying such a vast load of information? Pioneering research into the molecular structures underlying our genetic heritage go back to the early 1950s, when the biologist James Watson and the physicist Francis Crick realised that DNA forms a twisted string ladder, known as the double helix. The steps of this ladder are formed by pairs of DNA bases which fit together perfectly due to their complementary shapes and chemical details. The strings of the ladder consist of a repeating sequence of phosphoric acid and sugar molecules (Figure 8).

The base complementarity of DNA is a masterpiece of Nature, allowing the molecule to be copied with high efficiency. In this process,

Figure 8 *DNA*

it splits into two strands, each of which serves as a template for the construction of the missing half. When James Watson and Francis Crick published their famous double-helix model in 1953, they anticipated the existence of this mechanism and staked their claim at the end of the paper with one of the most famous understatements in the history of science: "It has not escaped our attention that the specific pairing we have postulated immediately suggests a possible copying mechanism for the genetic material."

Although the deciphering of the details of this mechanism and the underlying genetic code took many more years, Crick already described the DNA structure as a "code." As he pointed out in a letter, "if you have one row of letters, you can write the other one too. [...] The sequence of the letters (bases) distinguishes one gene from the next exactly like one printed page of a book is different from the other."

In contrast to human languages used in printed pages, the genes don't have any babylonian confusion of languages. Apart from a very small number of exceptions, their code is valid for all organisms. Its universality means that a given sequence of genetic "letters" is always translated into the same protein sequence. The discovery of this unity of Nature came as a surprise at first. Now, however, it is widely taken for granted as the foundation of today's gene technology. If, for example, a gene carries the instruction to make the human hormone insulin, it can not only be read in the human pancreas. Transferred into plant, yeast, or bacterial cells, it could direct the synthesis of the very same molecule in those cells just as well.

4.1 Why Does Michael have Dark Hair? Sugar and Phosphate Molecules Provide the Answer

As mentioned above, the DNA contains all the information necessary to "build" a human being. This information is lined up along the strands of the DNA molecules, organised in genes. Apart from a few genes specific for the stable RNA molecules needed in protein biosynthesis, every gene represents a protein which fulfils a specific task, for example in determining Bianca's eye and hair colour, or her blood group.

Each strand of the DNA double helix is a molecular chain, in which sugar building blocks alternate with phosphate groups. The fact that the sugar molecule has two different positions to which the phosphate groups can bind gives the chain a directionality. As Fischer explains: "In a chain of people holding hands and facing the same way, one can define a direction by saying the chain begins with a free left arm and ends with a free right arm. In a sugar phosphate chain, one can define the direction in the same way. The details of this definition are not important. What's important is that the two strands of the double helix always run in opposite directions."

The sugar building blocks also carry the DNA bases. There are four different kinds, which are labelled with the letters A,T, C and G. This simple alphabet is all that Nature needs to write our genes, and ultimately to build an organism. The available space and chemical details allow only two kinds of combinations between these bases: An A is always facing a T, and G is always paired with C. With this simple rule, one can conclude from the sequence of one strand which bases must be present in the other one, known as the complementary strand.

5 INHERITED GENES: A SURPRISE PRESENT

Building on these insights, researchers are now able to manipulate the blueprint contained in the genes. The identification, isolation and targeted editing of genes comprises the field of gene technology. It offers new, decisive impulses for the diagnosis and therapy of diseases, for reasons explained below.

Each (healthy) individual possesses 46 chromosomes: 23 from their father, and 23 from their mother. These chromosomes contain the DNA and thus the genetic information. The double set of 46 goes back to the fusion of a male with a female germ cell, each containing a single set of 23 chromosomes, resulting in the fertilised egg, which passes on the double set to all its daughter cells (except for the next generation of germ cells).

Thus, each chromosome can be paired up with a corresponding chromosome from the other parent. Therefore, one often refers to 23 pairs of chromosomes, even though, strictly speaking, only 44 of the 46 chromosomes in the human male form matching pairs. The exception to the rule are the sex chromosomes, which in the male come as a large X chromosome (from the egg) and a smaller Y chromosome (from the sperm).

This begs the question as to why most chromosomes (and all of the female) occur as matching pairs at all. One advantage of this arrangement is that if a gene on one chromosome has a defect, the fault can be compensated by the existence of a healthy copy on the chromosome of the other parent. If this compensation fails, an error in a single gene can easily lead to malfunction and disease, even if the other 30,000 genes are perfectly healthy.

There are two fundamentally different types of genetically caused disorders. In the first group are those that are exclusively due to a defect in one or several genes. For example, in the case of colour blindness, there is a clear link between the deficient gene and the disorder.

The case is more complicated for many widespread diseases, such as high blood pressure or cancer. While genes can play a decisive role in these, it is not always clear whether they are involved. In such cases, the gene alteration causes a genetic disposition for the disease, but not necessarily the disease itself.

CHAPTER 5

Hormones – The Body's Snailmail

"A drop of love is more than an ocean of determination and reason."
Blaise Pascal (French mathematician and philosopher, 1623–1662)

1 BLAME IT ON THE ADRENALINE

"Shall we go for a quick coffee?" Michael casts a questioning glance at Bianca. She nods: "After that overnight flight we should take it slowly – you don't have any other plans, do you?"

"No, of course not. Today you are the only point on my agenda."

"I like the sound of that." Bianca takes his arm. "Let's go to the 'Zeppelin' over there, that's where I had a sandwich earlier."

Because of the bulky luggage, the two choose a table at the periphery, and the friendly waiter Marco, who served Michael a bit earlier, is already approaching. He looks at Bianca and Michael. "So you found each other," he remarks in his Italian accent.

"Of course we have, and two cappuccinos please." "Coming straight away." Marco addresses Bianca, pointing at Michael. "Now he's much calmer. This morning he was really nervous, looking at the display all the time."

"Really?" Michael feels caught out. "Yes, that must have been the adrenaline," Marco adds laughing. "It's understandable though, it's due to *amore*."

1.1 Adrenaline and Co.: the Secret Rulers of Our Emotional World

With this remark, our friendly waiter has guided us towards another kind of messengers for the transmission of information in the body, namely the hormones. Like the neurotransmitters, hormones are chemical agents interacting with receptor molecules on nerve cells or other cells.

The difference between the two groups is that neurotransmitters act locally, across a synapse, while hormones can address all receptor molecules of a given kind, no matter where they are in the body. Because of this difference, hormones are much slower than neurotransmitters. While nerves can relay information in fractions of a second, the hormone communication can last minutes or even hours. When it comes to speed, nerves are the body's email, while hormones are more like snail mail. (see Box on p. 51)

Adrenaline and noradrenaline are the only fast-acting hormones. The brain evaluates very rapidly whether a situation is threatening. If the brain recognises danger, these two stress hormones are released. Travelling through the blood stream, they trigger a tightening of the muscles, and an increase of heart rate and breathing frequency. These physiological changes supply the body with extra power, and enable it to face the threat or to flee.

Chemically speaking, adrenaline is a hormone from the group of the catecholamines, which are synthesized in the adrenal medulla (the core of the adrenal glands, one of which sits on top of each kidney) and in a specific kind of nerve cells, the sympathetic ganglia. This group also includes the hormones noradrenaline and dopamine, which act similarly.

Adrenaline is released in situations involving physical and mental stress, infection, injury, and low levels of blood glucose, among others. By increasing the heart rate and blood pressure, widening the bronchi, promoting the oxygen uptake, and releasing fat and sugar reserves, it makes extra energy available (Figure 9).

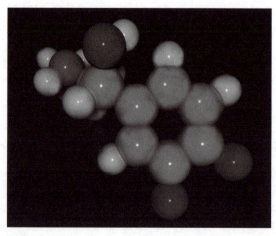

Figure 9   *Adrenaline*

HORMONE ESSENTIALS

Depending on their chemical structures, hormones are grouped into those that are mainly protein-like, and those that are more like lipids (fat). The former are known as peptide hormones and include insulin, glucagon, and the hormones of the pituitary gland and the interbrain (the middle section of the brain, also called the diencephalon). The fat-like group includes the steroid hormones, such as the sex hormones and the hormones of the adrenal cortex (the outer part of the adrenal gland).

According to their mode of action, hormones can be grouped in four categories:

- glandular hormones (classical hormones), which can act adenotropically (affecting other glands) or peripherically (affecting organs);
- aglandular hormones (tissue hormones), which are produced in specialized individual cells, which might be spread out over a given tissue, and are not concentrated in a gland structure;
- neurosecretory hormones (hormones of nerve cells); and
- mediators, which are hormone-like substances which can be produced and secreted by many cells, but only act locally because of their rapid degradation.

The hormonal system provides the organism with a second tool to control its internal environment, in addition to the vegetative nervous system. Hormones take care of the longer term and global control of cell functions, while the vegetative nervous system influences processes over a limited time.

"Adrenaline is an emergency hormone, which gives us the energy to survive by fight or flight," says the French physician Michel Odent, founder of the "Primal Health Research Centre" at London. As an expert for obstetrics and an advocate for a return to natural ways of giving birth, he has looked at the role of this hormone in the birth process of mammals. For example, if a mammal giving birth in the wild is threatened by a predator during labour, Odent says, the release of adrenaline stops and postpones the birth process, giving the animal time and energy to fight or flee.

The effect of adrenaline on the birth process is not a direct one, but rather more complex. During the last labour-pains before birth, both mother and baby experience the release of large amounts of hormones from the adrenaline family, making sure that both are wide awake after birth. For mammals in general, it is an advantage for the mother to have the energy to protect the newborn. For the babies, in turn, it means that they start life outside the womb with their eyes wide open. "Mothers are fascinated by the gaze of their newborn babies," says Odent.

He also thinks that this first eye contact is an important aspect of the nascent relationship between mother and baby. "It has to be emphasized that even the hormones of the adrenaline family, which are often seen as hormones of aggression, have a specific role to play for the interaction between mother and child in the hours after birth," says Odent.

Odent has applied the concept of "baby ejection reflex," which was originally used only for non-human mammals, in order to better understand the last labour-pains before the birth of a human baby. "This reflex only occurs, when the birth process can take place without disturbance," explains Odent. During this phase, women have the impulse to be upright, to grasp somebody or something, and they are full of energy. "Some women appear to be euphoric, others behave as though they were angry, others express existential fears of the transition," says Odent.

For all these behaviours, he has only one explanation: the sudden release of adrenaline. This triggers two or three strong contractions. "This reflex is rarely seen in hospital births, and even in home births if they are controlled by another person," says Odent, who started to practice alternative ways in the 1970s.

1.2 Hormone Deficiency in the Brain can Trigger a Rollercoaster of Emotions

The three hormones of the thyroid gland, too, can keep us on our toes. Apart from calcitonin, which plays an important role in the physiology of our bones, the two iodine-containing hormones, thyroxine and tri-iodothyronine, trigger an increase of energy turnover in all body cells, and can speed up cell division. The hormones of the thyroid also support growth and intellectual development.

Recent studies have revealed that these hormones have a much more important influence on mood than was previously thought. US scientists have developed a new therapeutic approach to treat pathological mood changes with the help of thyroid hormones (Figure 10).

Everybody goes through mood changes within certain limits. However, if these changes get out of hand in the long term, this may be due to a mood disorder. Around two percent of the population suffer from such phenomena, e.g. because they have manic-depressive disorder,

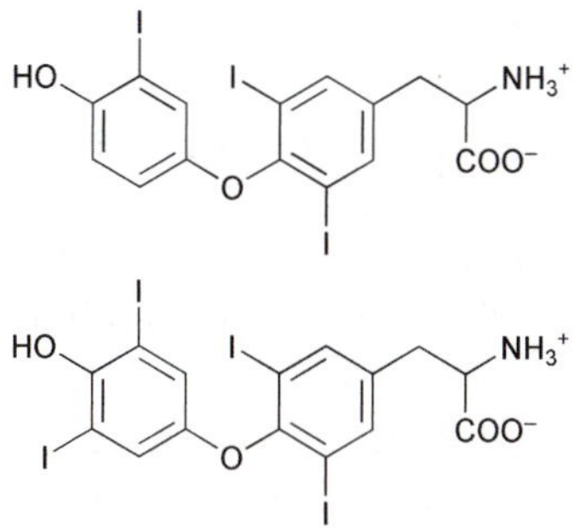

Figure 10   *Tri-iodothyronine (upper) and thyroxine (lower)*

also known as bipolar disorder. During the manic phase, they are awash with new ideas, spend money without consideration, and are generally uninhibited. After weeks or months, they fall into depression. Psychopharmaca and mood-stabilising drugs such as lithium salts can suppress the symptoms, but they also have unpleasant side effects and don't fight the cause of the disorder, which continues to exist beneath the surface.

"Conventional therapy fails particularly where phases are short and mood changes frequent," explains Peter Whybrow from the neuropsychiatric research institute of the University of Los Angeles. He has recently presented a therapy which can help these patients as well: administration of a mixture of the three thyroid hormones in high doses. "Some people will already improve after a single dosis corresponding to the natural hormone level in the blood. But most need a bit more," Whybrow reports.

His hunch that the brain might be suffering from hormone deficiency even if the concentration in the blood is normal led the researcher to the right track. The hormone molecule cannot pass the blood-brain barrier, so only its precursors can be imported into the brain. Moreover, some psychopharmaca and lithium interfere with the metabolism of the hormone, suggesting that these drugs would have undesirable effects.

Researchers had already suspected that thyroid hormones might influence the brain, as diseases of the thyroid gland often affect the mood. The breakthrough came with the application of positron emission tomography (PET), the brain imaging technique we mentioned in the first chapter.

Using this technique, researchers found that in manic-depressive patients the hormone activates the frontal lobes during the depression phase. These lobes control personality and intellect. In contrast, the limbic system, the seat of emotions, is suppressed.

1.3 Checks and Balances

In the short term, our mood may also depend on whether we are well-fed or hungry. This is the remit of another important hormone: insulin. It controls the concentration of glucose in the blood. Glucose, the product of partial degradation of carbohydrates such as sugar and starch, is an important fuel for all body cells.

Insulin is produced and secreted by a specific group of cells in the pancreas, the so-called beta cells, which are located within a structure known as Langerhans islets (the name "insulin" derives from the Latin word "insula" for island). Together with its antagonist, glucagon, insulin makes sure that the level of glucose in the blood stream remains within certain physiological limits. Glucagon acts to raise the glucose level, while insulin lowers it.

When a person ingests food, the beta cells respond in two ways. As soon as digestion starts, they release insulin from a reservoir into the blood stream. As this reservoir won't last forever, they also begin to produce new insulin, which will then be secreted slowly and continuously.

Insulin has a variety of effects in the body. Fat tissue, muscles, and the liver depend on its help for their uptake and use of glucose. It facilitates the import of glucose into the cell by stimulating certain structures in the cell wall. In addition, it activates enzymes in liver and muscle cells, which are in charge of the degradation of glucose. These processes are often referred to as "burning" glucose, as they liberate energy from a reaction with oxygen (Figure 11).

Thus, hormones control (almost) everything, including the metablism, which is the ensemble of all biological-chemical processes in the body. The hormone levels vary in the course of the day. This is why many people are more likely to be in a particular mood at a certain time of the day. Some even calculate their internal clock and plan the day according to their biological rhythms, while others dismiss this as fanciful.

Hormones, the body's information carriers, are produced in and secreted from the gland cells of certain organs. When they reach cells with the specific receptor molecules, the message encoded in their chemical structure can be read and acted upon.

Like all other human contacts, the relationship between Bianca and Michael is ultimately controlled by molecules. One might say that a sophisticated biochemical system is responsible for the various emotions of love. The sex hormone testosterone, which we will discuss in the next chapter, has a key role in this context, as even the libido, or sex drive, is essentially determined by chemistry.

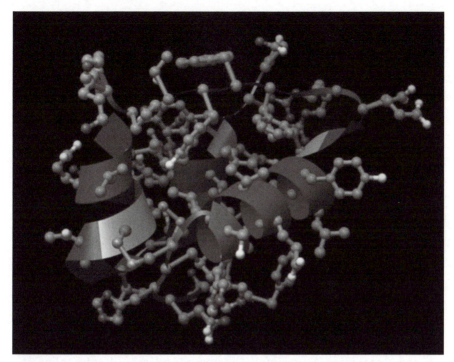

Figure 11 *Insulin*
©RCSB Protein Databank

No matter whether we are in love, happy, or sad: hormones always play an important role. They control many different processes in our body, from the blood glucose level, water management and blood pressure through to pregnancy and birth.

1.4 Exultation and Melancholy

As we have seen, hormones exert an important influence on our mental and physical health. During puberty and menopause, there are decisive changes in the hormone balance. During the developmental phase that turns children into adults, the hormones send us on a rollercoaster ride. The body changes. The moods fluctuate more violently, and the relationships with parents and friends change. Boys and girls develop an interest in their own sexuality. They enter their first relationships, experience first love. Due to the hormonal changes, the mood can oscillate between exultation and melancholy. During this time, youngsters need people they can talk to and who will stand by them.

The hormones, although they are just small molecules, wield considerable power. Not only do they make life exciting for teenagers, but they also take care of male and female fertility, and make the miracle of birth possible. They control and harmonise important processes in our body. The box below lists some of the most important hormones. Two of them, the male and female sex hormones will be the focus of the next chapter.

OVERVIEW OF THE MOST IMPORTANT HORMONES

- ACTH: short for adrenocorticotropic hormone. ACTH is produced in the anterior lobe of the pituitary gland and stimulates the adrenal cortex to form corticosteroid hormones. With this, ACTH contributes significantly to the control of metabolic processes and to the stress response. If the organism produces too little ACTH or none at all, the suprarenal gland will wither and produce insufficient amounts of cortisone. This hormonal disorder leads to a range of metabolic disfunctions.

- Adrenaline: As the name suggests, this hormone is produced in the adrenal glands, which are attached to the kidneys (thus *ad renal*). It increases the blood pressure and widens the vessels in the liver and in skeletal muscles. As adrenaline also stimulates the oxidative metabolism in the cell, its overall effect is an increased alertness of the organism. Accordingly, increased release of adrenaline is observed in stressful situations. When somebody is in love, this release, which happens within a few seconds, can be triggered by seemingly mundane situations, *e.g.* the lover appearing in the door, or the long-awaited ringing of the telephone.

- Androgenes: Collective term for the male sexual hormones produced in the testicles and in the adrenal gland. The most important androgen is the hormone testosterone, which is converted to androsteron and then excreted with the urine (see next chapter).

- Calcitonin: A peptide hormone that is produced in the so-called parafollicular C cells of the thyroid gland. Its release is triggered by an increased level of calcium ions in the blood. It slows down the continuous resorption of bone substance by the blood and stimulates bone formation, which also happens continuously. The combined effect leads to a reduction of calcium and phosphate levels in the blood. Calcitonin can be used against bone disorders such as osteoporosis.

- Cortisol: Also known as hydrocortisone. As the most important hormone of the adrenal gland, it enhances the degradation of food proteins, and the formation of storage carbohydrates, which will be stored in the liver. Its medical use is against allergies and rheumatism. It also has a pronounced anti-inflammatory effect.

- DHEA (Dehydroepiandrosterone): This hormone is one of the substances that have made the news in recent years, as there have been suggestions that it may prevent disease. According to some studies, the ageing process can be slowed down by topping up the natural DHEA levels. Therefore, the hormone is seen as a rejuvenating agent by some.
- Dopamine: As a so-called sympathomimetic, dopamine exerts a stimulating effect on the sympathetic nervous system. A lack of dopamine or dopamine receptors in the brain leads to Parkinson's disease. Dopamine can also make you euphoric and acts as a happiness hormone for the love birds.
- Endorphines: These brain or neuropeptides have a dual function as hormones and neuropeptides. They bind to opiate receptors in the brain and spinal cord, thus blocking the communication of pain. They will be described in detail in Chapter 13.
- Estradiol: Sex hormone from the group of - > estrogens (see there).
- Estrogens: Collective name for the female sex hormones produced in the follicles of the ovary. Other production sites include the adrenal glands, the fat tissues and the male gonads, as estrogens occur in the male organism in small quantities as antagonists to the androgens. In general, estrogens control the normal sexual cycle together with other hormones. In women, the estrogen level is particularly high at the time of ovulation, which has a positive influence on their ability to experience orgasm. For men, in contrast, estrogen excesses will reduce their libido.
- Follicle stimulating hormone (FSH): Also known as follitropin. This glycoprotein of the anterior lobe of the pituitary gland supports growth and development of the gonads and stimulates them to produce estrogen. In women, it plays an important role in the menstrual cycle, as it triggers the maturation of a new egg and its production of estradiol. In the male gonads, too, the FSH fulfils essential functions, as it stimulates the production and maturation of sperms.
- Insulin: Pancreas hormone produced in the beta cells of the Langerhans islets. Together with its antagonist glucagon, insulin maintains the blood glucose level within certain limits. Glucagon raises the glucose level, while insulin lowers it. A lack of insulin triggers diabetes.
- Luteinizing hormone (LH): A hormone produced in the brain, which triggers ovulation and the production of the sex hormone progesteron in women. In men, it stimulates the production of the sex hormone testosterone.

- Melatonin: This melanin-concentrating hormone is produced in the pineal gland. It inhibits the function of the thyroid gland and lowers the secretion of luteinizing hormone. Thereby it blocks sexual maturation processes and generally slows down metabolism. By releasing melatonin, the pineal gland controls the day night rhythm of the body. In daylight, the release of the hormone is blocked. After dark, melatonin is released and exerts its sleep-enhancing effects. Travellers across time zones and shift workers experience a mismatch of the melatonin cycle with their sleeping pattern, a phenomenon generally known as jet lag. Taking melatonin at exactly the right time can help to avoid jet lag.
- Noradrenaline: This compound is physiologically related to adrenaline. It is produced (together with adrenaline) in the adrenal medulla as a hormone, but also in the brain, in the area of the sympathetic ganglia, as a neurotransmitter. Its physiological effect is to increase the blood pressure by narrowing the peripheral blood vessels, which is thought to be mediated by the opening of calcium channels. In addition, it steadies the smooth musculature while stimulating the heart muscle.
- Oxytocin: Produced in the posterior lobe of the pituitary gland, this hormone stimulates the labour constrictions of the uterus and the function of the mammary glands. As a "cuddly hormone" it is sexually stimulating and is thought to be responsible (at least partially) for the longing for closeness and comfort.
- Phenyl ethylamine (PEA): Or beta-phenyl ethylamine, to be precise. This relatively simple molecule is classified as a biogenic amine and occurs in the brain, among other organs. As the core structure of various hallucigenic substances, PEA is connected with feelings of desire and happiness. Small wonder, then, that it is also found in chocolate.
- Progesterone: The ovaries produce this hormone during the second half of the menstrual cycle, while the lining of the uterus is being prepared to accommodate the fertilised egg. If a pregnancy is established, the placenta will continue to produce this hormone. Progesterone inhibits the maturation of any further follicles, reduces the sensitivity of the uterus musculature, and stimulates the production of milk in the mammary glands.
- Prolactin: Hormone produced in the anterior lobe of the pituitary gland, which stimulates the tissue proliferation of the mammary glands during pregnancy and the secretion of milk (lactation) after birth.

- Serotonin: Not a hormone, but a neurotransmitter. However, we have listed it here because it is sometimes referred to as a "happiness hormone," like dopamine. The healthy human organism contains around 10 mg serotonin. A reduction of this amount leads to a mood swing, with symptoms like lack of drive, sleep disorder, anxiety and depression.

- Testosterone: The most important male sex hormone is responsible for the formation of the secondary sex characteristics of the male. It supports the development of muscles, bones and red blood cells. It is also involved in the growth of body and facial hair, and in other phenomena that kick in with puberty, including the growth of penis, testicles, and the prostate. Finally, it is important for the maintenance of virility and libido. – Testosterone is also found in women's blood. Although the concentration is around five times lower than in men, some researchers suspect that the corresponding testosterone receptors are more sensitive than those in men. This might explain why small fluctuations of women's testosterone concentration have stronger effects on psyche and desire than they would have in men.

- Thyroxine: The most important among the hormones of the thyroid gland, which also include tri-iodothyronine. The thyroid hormones are responsible for a variety of metabolic functions. Their overproduction increases the heart rate and the sensitivity of the nervous system. A lack of these hormones, in contrast, can interfere with the intellectual development of children and lead to a general loss of drive, or even to apathy in adults.

- Vasopressin: This hormone is produced in the occipital lobe of the pituitary gland and controls the function of the kidneys. A lack of vasopressin triggers a massive overproduction of urine (diabetes insipidus). Like oxytocin, vasopressin is also involved in our emotional life.

- Growth hormone (somatotropin): This hormone is produced in the anterior lobe of the pituitary gland. Together with other hormones including insulin and thyroxine, it is responsible for body growth. A lack of this hormone can lead to dwarfism, but this can now be compensated by administration of appropriate doses.

CHAPTER 6

Estrogen and Testosterone – Driving Our Emotional Lives

I will reveal to you a love potion
without medicine, without herbs,
without any witches' magic
if you want to be loved, then love.
Hecaton of Rhodes (Greek philosopher, ~100 BC)

1 THE SECRET ENTANGLEMENT OF BODY AND SOUL

The phenomena of love and passion have interested people since the dawn of history. Ancient cultures didn't have the slightest idea of the complex chemical processes that occur in the body, driving these phenomena. Thus, the topic offered fertile ground for myths and legends.

In ancient Greece, there was a myth providing an explanation for the overwhelming love which two people might feel for each other. The story went that two such people had – in an earlier life – been one person, a strong and high-spirited being. This met the dislike of Zeus, who cut the being in two. Since then, the two parts run around the world, driven by the ardent desire for their missing half. When they find each other, they are overjoyed.

We love this old story, because it describes in a mythical way the intimate entanglement between man and woman, body and soul, an entanglement that can be frightening and tragic, often incomprehensible, but it is in our nature, and it is always enchanting.

2 BIANCA KNOWS THE STORY BUT CAN'T FIGHT HER FEELINGS

Bianca and Michael cannot withdraw from this entanglement either. As "enlightened" people, they know the underlying mechanisms, the interplay of the hormones, which sometimes can play tricks on them.

Thus, Bianca always goes through a low mood during the days before her menstruation. Even though, as a medical student, she knows the physiological context full well, she just cannot escape the downhill slide of her emotions. Fortunately, Michael is a very understanding partner, who treats her with special care during such phases, and often just takes her in his arms and caresses her.

The main responsibility for the emotional rollercoaster that women experience in their menstrual cycle lies with the female sex hormones, the estrogens and progesterone. The release of the sex hormones is controlled by the so-calleld releasing hormones of the hypothalamus. These are polypeptides consisting of only ten amino acids, which are released at regular intervals throughout the day. These relatively small peptides stimulate the anterior lobe of the pituitary gland to produce a range of stimulating hormones.

The most important releasing factor for women is the gonadotropin releasing hormone. Hormone researchers now generally agree that at the beginning of the menstrual cycle an as yet unidentified part of the brain sends signals to the hypothalamus. Some specialised neurons of the hypothalamus produce the gonadotropin releasing hormone, which is then released in regular pulses into the blood vessels leading to the pituitary gland.

As a recipient of these hormonal cascades, the pituitary gland then produces follicle-stimulating hormone (FSH) and luteinising hormone (LH). FSH in particular stimulates the growth of follicles in the ovary. The follicle, a kind of bubble that envelopes the egg cell, grows and becomes a small reactor which produces the estrogen estradiol from cholesterol. A few days later, the events have progressed to the point where the pituitary gland releases LH, which in turn triggers ovulation.

However, the task of the follicle in the cycle is not yet completed. It now turns into a corpus luteum, which produces the gestagen progesterone. This prepares the lining of the uterus for the possible implantation of a fertilised egg. If that doesn't occur, an abrupt drop in the levels of estradiol and progesterone at the end of the cycle leads to menstruation. During the entire cycle, the appropriate concentration of hormones in the blood is regulated by a complex system of feedback controls.

The progesterone level in the blood also depends on the phase of the menstrual cycle and can undergo enormous fluctuations. Progesterone dominates the second half of the cycle. It prepares the uterus for a possible pregnancy. If a pregnancy is indeed established, the hormone allows it to continue, prepares the production and secretion of milk, and permanently increases the basal temperature. The basal temperature is the body temperature measured in the morning, directly after waking up. While a woman is not pregnant, it increases after ovulation by half a degree and drops back to the initial value just before menstruation.

2.1 During Puberty, the Countdown of the Body's Chemistry Begins

The invisible chemistry which occurs in our bodies is the secret ruler of our emotions and our entire personality. During our childhood it remains asleep, waiting for the right time to unfold its power.

During puberty, estrogens trigger the formation of the typical female sex characteristics, including:

• the development of the rounded breasts,
• the establishment of the "high" adult female voice,
• the manifestation of typically female patterns of hair growth and distribution of fat tissues creating the "curves."

Estrogens also stimulate the maturation of bones, lower the cholesterol level and increase the incorporation of water in tissues. Medically, they are used as part of the birth control pill, for the therapy of menopausal disorders, and for the treatment of osteoporosis (Figure 12) We will come back to the hormonal therapy of problems associated with the menopause in Chapter 12.

Synthetic hormones which are similar to progesterone are known as gestagenes. They are used for birth control and for the therapy of some hormone producing tumours.

2.2 Men are Much Less Complicated, or are They?

In men, the control of sex hormones follows somewhat simpler rules. Here, too, gonadotropin-releasing hormone acts as a trigger. Together with FH and LH from the pituitary gland, it controls the production of the male sex hormone testosterone in the testicles. Testosterone is responsible for the

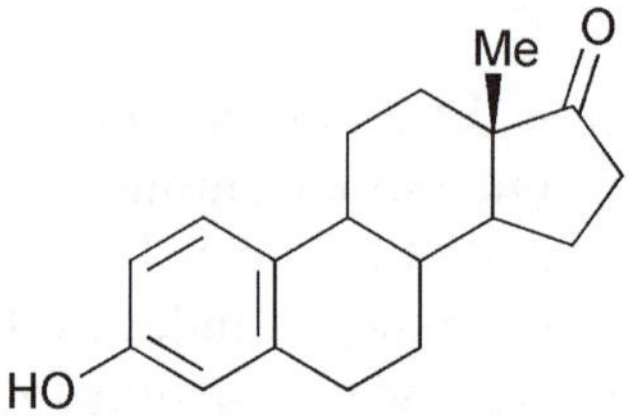

Figure 12 *Estrogen*

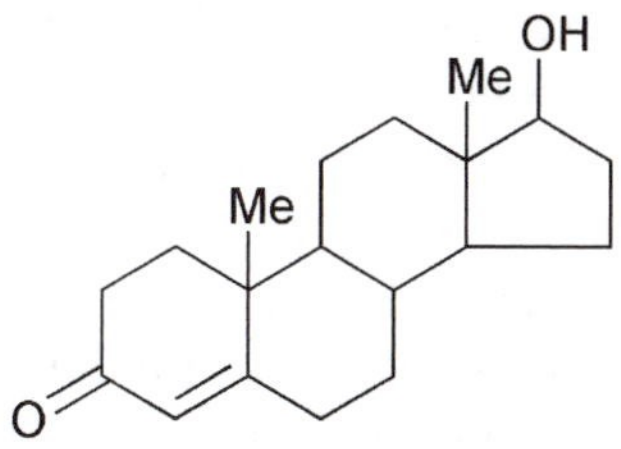

Figure 13 *Testosterone*

maturation of sperm cells, but it also has a feedback effect on the brain, where it can influence different groups of nerves (Figure 13).

Additionally, small amounts of both female and male sex hormones are produced in the adrenal glands and in the liver. "Nevertheless, a normal male or female sexual behaviour will always depend on an intact axis of hypothalamus, pituitary gland, and gonads," as biologist Gaby Miketta explains. The mechanism depends not only on the sex hormones, but also involves various neuropeptides, as researchers have found recently.

For boys as for girls, puberty begins with an increased production of the specific sex hormone, *i.e.* testosterone in the case of boys. The increase of the testosterone level causes the development of the male sex organs. Testicles, scrotum, penis, prostate, epididymis, and seminal vesicle grow disproportionally. The first ejaculation is most likely to happen in the 14th year, typically at night and spontaneously.

During puberty, the typical male hair pattern develops. In fact, the growth of pubic and underarm hair is one of the first signs of puberty. In a later phase, the pubic hair spreads in a diamond shape towards the navel, while growth of chest and facial hair kicks in.

The Growth of the Larynx –a.k.a. "Adam's apple" – leads to a displacement of the vocal cords. This development makes the boy's voice "break," resulting in the typical deep male voice.

As a natural anabolic, testosterone encourages protein synthesis, which leads to a marked increase in muscle mass. The resulting muscular body shape is the counterpart to the female curves. During puberty, there is a burst in growth, which for boys is fastest between the 14th and the 16th year.

With the end of puberty, the final body height is reached. The growth process, too, is controlled by testosterone. The hormone stops growth by ossifying the cartilage zones at the ends of the tubular bones.

Testosterone also exerts a crucial influence on sexual desire (libido). Thus, the hormone does not only determine male fertility, but also virility.

This connection is exemplified by the spontaneous erections in the morning. Testosterone concentration in the blood undergoes fluctuations throughout the day, and the highest values are reached in the morning.

Testosterone is indispensable for the formation of sperms in the testicles and thus has a direct effect on fertility. The frequency and quality of orgasm also depends on the effects of testosterone. Many men claim that the amount of sperm ejaculated is an indicator of the quality of their orgasm.

From the viewpoint of biology, this might well be true. As testosterone influences the function of the seminal vesicle and prostate, the amount of sperm produced will of course depend on testosterone levels. However, we doubt whether the quality of an encounter can ultimately be measured in millilitres, as there is also a mental component which may decide how satisfactory the experience was for both sides.

What remains certain is that testosterone has a crucial influence on the quality of sperms. There is a simple reason for this, which has to do with the chemistry of the ejaculated fluid. Among other things, the hormone controls the fructose content of the fluid, which in turn is crucial for the energy supply of the sperms, and thus for their ability to perform.

2.3 Testosterone – A Versatile Hormone

There are many more things that testosterone can do. Without exaggeration we can say that the male sex hormone is a Jack of all trades. Thus it has a unique way of looking after the entire male body image and the performance of the male organism. Not only does it determine the pattern of hair growth, it also stimulates the activity of the sebaceous glands and thereby controls the fat content of the skin.

The build-up of muscle mass, and thus the force of one's muscles also depends on testosterone. The hormone not only enables the formation of new muscle cells, but also the growth of existing fibres. As many male teenagers will be happy to demonstrate, this increase in muscle mass also brings about an increase in their sheer force.

Testosterone is also reponsible for the ratio between muscle and fat tissues. A lack of testosterone leads to a replacement of muscle tissue with fat tissue. When men over 40 begin to show their age and increase in weight, the slowly waning production of testosterone has made its mark. However, this does not mean that every beer belly is an inevitable result of the changes in body chemistry.

DOES MASTURBATION PREVENT PROSTATE CANCER?

An apple a day keeps the doctor away, as the old saying goes. According to a study from the Australian urologist Graham Giles of the Cancer Council Victoria at Melbourne, there may be other routines with the same effect. When "New Scientist" picked up his story, this must have caused astonishment in some and sheer horror in others. "Can masturbation keep the doctor away?" the headline ran. One can imagine the reaction.

Giles is unfazed, however. He had questioned 1079 men who had prostate cancer about their sexual habits, and compared the results with a control group of 1259 men who had no problems with their prostate.

From the answers he calculated that masturbation may reduce the risk of prostate cancer by up to one third. Regular practice between the ages of 20 and 30 was found to have a particularly protective effect.

Until then, medics had assumed that sex was bad for the prostate, as earlier studies connected frequent intercourse to a 40% rise in the risk of prostate cancer. Giles and his colleagues suspect that this rise is due to the exposure of the prostate to infectious agents during intercourse, a danger that can be ruled out in solitary pleasures.

However, the exact causes of any protective effect that masturbation may have remain open to speculation. It might be possible, the researchers suggest, that ejaculation removes cancerogenic agents from the body. The prostate is known to enrich certain substances from the blood, in order to provide the sperms with potassium, zinc and citrate. By the same mechanisms, harmful substances could also accumulate in the prostate, unless you have the habit of releasing them quite regularly.

Indirectly, testosterone also contributes to the energy budget of the body. The hormone stimulates the production of red blood cells, the blood's oxygen carriers. As oxygen is an important requirement for the body's production of energy, all our metabolism depends on it.

A few years ago, it was established that testosterone also stabilises the bones and helps to maintain the bone density. Furthermore, it plays a role for the general well-being, the mental ability, and the drive and mood. A lack of testosterone in men can lead to various ailments and even to serious disorders. For young men, the spectrum of the consequences includes failure of the voice to break, lack of muscle formation and body hair, through to social immaturity.

Talking about immaturity, we have noted with slight irritation a study from an Australian researcher, which we present and comment in the box on page 65.

Did that make you smile? Good. Of course one can speculate about everything, including safer sex, but maybe it would be premature to base any medical recommendations on this result. If anybody does want to try this at home, we recommend the high tech option, replacing old-fashioned masturbation with cyber sex.

To return to our topic, which normally involves two persons, rather than one, we have always felt intuitively that a harmonic marital life is an important factor contributing to good health. And lo and behold, there are new studies from British and US researchers showing that our intuition wasn't totally wrong. Even though the researchers did not specifically include tumour rates, the results appear to be clear.

3   LOVE IS THE BEST MEDICINE

We always knew that, didn't we? A life in peace and harmony, while no guarantee for eternal health, is a solid defence against all those disorders that are more or less caused by unhealthy stress. It should be obvious that this defence should also include a healthy sex life.

The magazine "Men's Health" stated that "sex is the best health provision." According to the magazine, regular sex can make the physician or pharmacist unnecessary. This headline-grabbing claim was based on a long-term study conducted at the University of Bristol. The researchers conclude that at least for men there is a direct link between sexual activity and health. Sadly, it remains unclear whether women benefit as well.

According to the Bristol researchers, the effects of healthy sex in men range from higher resistance against the common cold, through to the prevention of heart disease and cancer. Furthermore, the increased testosterone level is thought to improve the memory capacity, and reduce the risk of a stroke.

Sex is also thought to be a good painkiller, as the opiate-like substances produced by the body (see Chapter 13) can relieve joint and head aches. The messenger dopamine brushes aside all perceptions of stress for about two hours. Various hormones that are released in increased amounts during sex boost the immune system, protect from arteriosclerosis and osteoporosis, and strengthen both muscles and bones.

Independently, the research team and married couple Janice Kiecolt-Glaser and Ronald Glaser at Ohio State University achieved similar results. As befits a couple, the Glasers have investigated specimens of both sexes. After 20 years of studying married couples, they claim that the quality of a marriage has a positive influence on both partners' health. In a nutshell: In a happy relationship, the partners are healthier.

3.1 Domestic Disputes Weaken the Immune System

The scientific evidence of these researchers is based on long term measurements of the levels of stress hormones and the wound healing process in 90 couples who were newly married at the beginning of the study. It emerged that both men and women respond to the quality of their relationship via the level of stress hormones in the blood, and the strength of the immune functions. The researchers found, for example, that a serious domestic dispute significantly weakened the immune system. As a consequence, both the efficiency of vaccinations and the times required for wounds to heal deteriorated.

In order to mimic a wound, the researchers exposed the participants of their study to a vacuum on their arms, resulting in a blister. "We were able to record quite precisely what happened in the wound during a social interaction," Ronald Glaser explains. Specifically, the researchers focussed on cytokines and on neutrophiles, a special kind of white blood cells. Stress in the relationship, they say, can lead to a retardation of the healing process via a blockage of the cytokines.

In divorced couples, the researchers found conspicuously high values of stress hormones such as adrenaline, corticotropin, and cortisol, in contrast to the couples who lived in a harmonic relationship or at least tried to.

"Even the fact that married couples talk about changes – be it that they want to get more organised, or that they plan to lose weight together – generally has a positive effect on their health," says Glaser. A positive attitude is also reflected in reduced levels of cortisol. Glaser concludes: "The lower the cortisol level, the faster at least the external wounds can heal."

CHAPTER 7

Oxytocin – the "Amuse-Gueule" Among Hormones

"To love does not mean to look at one another, but to look in the same direction."
Antoine de Saint-Exupéry (Writer, 1900–1944)

1 WHEN MICHAEL CARESSED BIANCA, HER "CHEMICAL FACTORY" PRODUCED OXYTOCIN

Like many relationships, the one between Bianca and Michael is not only of a sexual nature. For the depth of the relationship, the desire for tenderness, comfort, along with physical and mental closeness of the other part is more important – emotions that go beyond the mere physical attraction.

It may sound disillusioning, but for this level of emotions, too, chemistry has decisive responsibility. Researchers have found out by using animal experiments that chemical substances in the brain account for the feeling of affection. In rats and mice, for instance, this feeling expresses itself by snuggling up to each other and cuddling. "In mice it's oxytocin, but in rats the related hormone vasopressin might be involved in this process," reports hormone researcher Richard Ivell at the University of Melbourne, Australia.

Ivell also believes that molecules like oxytocin and vasopressin have accompanied life as essential messengers for partnership and faithfulness from very early on. "Oxytocin is an old hormone in evolutionary terms, as it is also found in simple animals like earthworms," he emphasizes.

"For the relationships between human beings there appears to be a molecular connection," Ivell adds. All experiments so far have suggested that this connection is made primarily by oxytocin. This opinion is shared by the US scientist and widely known author, Helen Fisher, a biological anthropologist at Rutgers University, New Jersey. Together with colleagues at the Albert Einstein College of Medicine at Stony Brook, she is pursuing a research project dealing with the "brain physiology of romantic love."

In the course of this project, she places newly enamoured people in the rather unromantic tube of an MRI (magnetic resonance imaging) scanner. "I want to read from their brains how they love," Fisher says. We have already met the MRI technique, which enables her to read the brain activity, in the first chapter. It makes particularly active areas of the brain visible because of their increased oxygen flow.

In order to track down the secrets of romantic love, Fisher showed two photos to the volunteers. First one of their lover, then one of a "neutral" person. In both cases it is to be expected that certain areas of the brain become active. "If we subtract the areas that have responded to neutral people from those that have responded to the loved person, we should be able to see exactly which parts of the brain are active when somebody is in love," Fisher hopes.

On the basis of previous studies, she concludes that a specific chemical system is responsible for the stimulation of different brain regions, and thus for the different feelings of love. "We know that libido in both men and women is largely controlled by the hormones testosterone and estrogen," Fisher explains. The close connectedness, another decisive emotion, is connected with the hormones oxytocin and vasopressin. "These are the substances in the brain that transmit the feeling of deep affection," says Fisher.

1.1   Keen on Cuddling after Orgasm? It's all in the Chemistry!

As a love and feelgood hormone, oxytocin becomes activated on a large scale after tender or pleasurable touch, but especially during sexual arousal. The brain produces a strong burst of the hormone after orgasm, thus creating a feeling of safety and comfort. Pleasant closeness and familiarity are the dominant emotions. It might be due to the oxytocin that relationship problems often vanish after a successful amorous encounter.

Some men wonder why women are keen to cuddle up after the sexual climax. It's quite simple: The oxytocin released during orgasm plunges them into their own world of intensive emotions, a paradise of maximal closeness and connectedness, where they want to remain as long as possible. Thus, oxytocin can be seen as a self-produced drug with a strong euphorising effect (Figure 14).

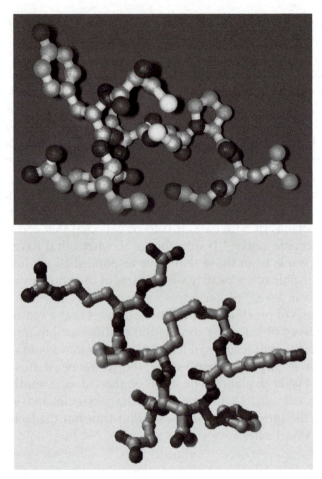

Figure 14   *Oxytocin (upper) and vasopressin (lower)*

1.2   From Moscow with Love: Oxytocin and the Pavlov Effect

When the psychologist Diane Witt of Binghamton University in the state of New York describes the effects of oxytocin, it sounds almost as though she's reminiscing about her own teenage years. "First you find a boy cute. Then you go round the block with him, and you start finding him extremely cute. But when you finally share a bed with him and experience an orgasm, then you cannot imagine a life without him," as she describes the effects of the cuddle hormone.

Witt even attributes a certain Pavlovian effect to the sex hormone. This requires a short explanation. The effect is named after the Russian

Figure 15   *Ivan Petrovich Pavlov*

Nobel laureate (medicine, 1904) Ivan Petrovich Pavlov (Figure 15). Around the turn of the century, in the course of his empirical studies, which are today seen as a foundation of both ethology in animals and human psychology, he had conducted an experiment with dogs, which soon became universally known as "classical conditioning."

In his experiments, Pavlov had combined the provision of food with a bell ringing. When the animals could see and smell the food, the bell rang. After only a few rounds of conditioning, the acoustic signal alone was sufficient to make the dogs salivate to the same (measurable) extent as the sight of food. The bell turned into a conditioned signal, and the reaction into a learned, *i.e.* conditioned reflex.

"If you have sex with the same partner several times, this can lead to the situation that the oxytocin level rises simply from looking at the partner," Witt claims. Which of course would lead directly to the desire to proceed to the bedroom with the person in question.

Thus, the hormone is rightly considered a long-term glue for relationships. However, Fisher suspects that there may be a less appealing flip side to this coin. "I assume that pathological relationships may also be caused by oxytocin," she says. An overdose of the hormone might trigger an exaggerated mother-child bond. It might also be possible that infatuation with or sexual dependence on a partner could be due to an overdose of this glue.

Other important functions of the hormone are in the area of breast-feeding. An oxytocin containing spray known as "Syntocinon" is already available on prescription for young mothers. It stimulates the steady milk production and prevents damage to the nipples. Via the internet pharmacies, the product also finds its way to a more surprising group of customers: show girls. They like to apply small doses of the hormone to make their nipples perk up.

But still, this is far from being the end of the range of effects that oxytocin exerts in the body. In men it triggers rhythmic contraction of the seminal duct. In women giving birth it stimulates the contractions of the uterus. Obstetricians have long used the hormone as a means to support labour. In the US, synthetic oxytocin is used in three out of four hospital births to ensure a rapid birth. Which makes sense, as this is exactly what the Greek name suggests.

By now, researchers have found out that one doesn't have to be in love in order to let the hormone flow. Caresses and tender massages are sufficient. Such a gentle touch can beat the blues and they are a treat for body and soul. Intuitively, Michael did the right thing when Bianca was down and he just caressed her tenderly, without a hidden agenda.

1.3   Why Women Pick up the Phone When They are Stressed

According to recent research, oxytocin is also responsible for a wide-spread phenomenon – that women are better able to cope with stress than men. Researchers at the University of California at Los Angeles (UCLA) have investigated why this might be the case.

The project was led by the psychologist Shelley Taylor, a stress researcher who focusses on the interplay between biology and human behaviour. It revealed that oxytocin plays an important role in stress situations and is important for the gender-specific ways of coping with stress. Taylor and her colleagues have observed that women in stressful situations intuitively establish contacts to other people. "They make phone calls, they converse, they communicate – 'tend and befriend', as we call it," she explains.

According to the study, it looks as though specifically female hormone reactions play a role in this behaviour. Taylor comes to the point: "Under stress we release a whole host of hormones, especially oxytocin. The effect of oxytocin is enhanced by estrogens, of which women have a lot. Oxytocin is a bonding hormone, so it establishes bonds with other people, and it is also produced upon contact with others. It is a very positive hormone, so to speak. Some people here call it the cuddle hormone."

This cuddle hormone is generally known for being released in large quantities during breast feeding. Which, again, has a lot to do with contact. "Obviously, oxytocin plays an important role in all activities related to reproduction," hormone researcher Ivell explains. For example when the birth of a child approaches, oxytocin is released and triggers the contractions of the uterus, which in turn mark the onset of labour. It is similarly involved in ovulation and milk production.

"Many mothers report a peculiar kind of happiness experienced during breastfeeding," Ivell emphasizes. The harmony between mother and baby, too, is controlled by oxytocin. Recently, researchers have shown that the oxytocin level of a mother increases when her baby is signalling hunger. This can be seen as a parallel to sexual arousal, which often increases even before any stimulation of the skin has taken place. When a baby suckles, the level of oxytocin released by the mother is about as high as during orgasm – another parallel between these two central events in our reproductive life. Oxytocin is also contained in breast milk. Thus, the baby absorbs a certain amount of the love hormone through the digestive tract.

When Bianca and Michael fell into each other's arms at the airport, it must have been the influence of the oxytocin produced by their body's chemistry which made their blood boil and made them experience an euphoric feeling of warmth and comfort. Ivan Petrovich Pavlov sends his regards...

1.4 Small Molecule, Big Effect

Even Ivell, an experienced hormone researcher, finds it remarkable that oxytocin can trigger this kind of reaction. "It is a remarkably small molecule," he states. Biochemically speaking, oxytocin is a nonapeptide, *i.e.* a molecular chain consisting of nine amino acid building blocks, six of which are closed to a ring via a disulphide bridge. It is produced in the hypothalamus, from where it moves to the posterior lobe of the pituitary gland, which will release it upon demand.

Until very recently, oxytocin was assumed to be an exclusively female hormone with the tasks of triggering labour and lactation. By now, however, it is seen as a male and female hormone involved in all aspects of love. Researchers at the Max-Planck-Institute for Psychiatry at Munich have found that in men, too, it helps coping with stress. According to Shelley Taylor, however, the positive effect that oxytocin has in women is compensated in men by the activity of their sex hormones, especially testosterone.

The consequence is that men respond to stress either by aggressively fighting the perceived source of it, or, if the situation appears hopeless, by trying to flee from it. The generalisation of this behaviour became a dogmatic view, which was to block research for a long time. Over decades, stress researchers were convinced that the "fight or flight" pattern was the typical response of human beings, regardless of their gender. Previous studies had shown that a complex hormonal reaction in the body could explain this classical form of aggressive or evasive behaviour.

It was easy to construct a historical context, as researchers unanimously believed that we inherited this behaviour from our ancestors in the stone age. Especially the early humans, as the convincing theory went, could best save their families from threatening situations by seeking fight or flight. Animal experiments appeared to confirm this assumption.

However, Shelley Taylor proposed a different hypothesis. She had noticed that the behaviour of many women just didn't match this stereotype. Moreover, she discovered a few contradictions in the previous studies.

She realised that all animal experiments had been conducted exclusively with male animals. Among the human subjects, more than two thirds were men, and there had been no systematic studies to check for possible differences between men and women. Thus, the fundamental models of stress research were based on studies that had failed to take gender factors into account.

The latest insights from Taylor and her group are based on the analysis of hundreds of studies from around the world and many experiments from her own laboratory, addressing the role of hormonal changes.

From the angle of human prehistory, it makes sense to look for a difference between the stress response of men and women. "Stone age women reacted to a threat by "tending and befriending", they looked after the children and got organised together with the other keepers of the fire," Taylor argues. For the survival of the clan, there was no point in leaving the children alone and follow the men into battle.

"In the modern world, similar behaviours can be observed," Taylor continues. Studies from Scandinavia, for example, had shown that men had a desire to be left in peace after a stressful day at work. Taylor: "Some did sports, others wanted to watch TV, and a third group reacted aggressively." Many mothers, on the other hand, established contact with their children, or talked to their friends, dealing with the stress via human contact.

1.5 Men and Women, Be Tolerant!

"Moreover, the 'tend-and-befriend' -behaviour is healthy," Taylor argues. From a multitude of studies it becomes clear that people who have many friends and acquaintances live longer, are less frequently ill, and recover more quickly. This holds not only for women, but for men, too.

Based on her studies, Taylor would like to offer a piece of advice to all couples: "It would be healthy for the coexistence of the sexes, if men could develop a certain tolerance towards the communication-intensive behaviour of women," she says. On the other hand, women should also be aware that the male "fight or flight"-response is a natural reaction to stress.

To return to the hormone responsible for all that tending and befriending, a few researchers set out to investigate the precise locations where oxytocin is released in the human body. They suspected that the brain might not be the only source of the hormone. To answer this question, they had to search all over the body, by taking tissue samples, dying them with specific solutions, and studying them under the microscope. "In fact we found out that the hormone is also produced in the ovaries, the prostate and the testicles," Ivell reports.

Ivell has also tried to address the more difficult question in which way oxytocin exerts its influence on our emotional world. For this, tissue samples are no use. Instead, Ivell went for an indirect method. Working with a group of male British student volunteers, he investigated the level of oxytocin in the blood before and after masturbation. As was to be expected, the level increased to several times the initial value after orgasm in all participants. "There was a real burst of oxytocin release into the blood," Ivell explains.

The surprising and crucial result, however, was obtained with the same group in a second round of experiments. Once more, the volunteers were asked to masturbate in the service of science. However, this time they were given an oxytocin blocker beforehand – a drug which suppresses the production of the hormone. When the students were asked later, whether they noticed a difference, they all reported that everything was normal physically, but that this time they had gained no pleasure from the procedure.

The US psychologist Mary Carmichael performed the matching experiment with female volunteers at Stanford University. Monitoring the oxytocin levels during orgasm of men and women using an intravenous catheter. Surprisingly, she found that the level during masturbation is much higher in women than in men. Moreover, women who were capable of multiple orgasms achieved an even higher level the second time round.

1.6   Social Behaviour Benefits, Too

Does that mean that oxytocin is a kind of amuse-gueule, an appetiser among hormones? "Yes, it is, but not only that," Ivell clarifies. Oxytocin is a remarkable molecule, involved in everything to do with relationships.

There have been countless experiments demonstrating the effects of oxytocin in animals. If for instance chickens or pigeons receive an injection of the hormone, they start display and mating within a minute. Therefore, breeders have used oxytocin for decades to get animals to mate.

A surprising social component of the hormone effect becomes apparent when it is injected into the brains of virgin rats. If researchers presented rat-like decoys to these animals after the injection, female rats began to care for the decoys and behave like mothers.

The exact opposite effect could be observed when mother rats received an antagonist to oxytocin shortly after giving birth. The animals gave up their nests and no longer cared for their offspring.

The hormone also has a crucial effect on the bonding between couples. This time, the experiments were not conducted with rats, but with prairie voles. These animals form permanent partnerships, which is relatively rare among comparable animal species. US researchers have shown that the prairie vole, compared to other, promiscuous species of vole, has a different distribution of oxytocin in the brain.

1.7   Oxytocin and Vasopressin – Chemical Compass Needles for Partnership and Fidelity

Such permanent bonds are also the focus of the research of Thomas Insel, a neurologist at Emory University in Atlanta, Georgia. In order to reveal the secret of the prairie voles (*Microtus ochrogaster*), Insel compared them with the genetically related montane voles (*Microtus montanus*), which do not form any bonds but copulate at random. The investigation showed that the "faithful" prairie voles produced significantly larger amounts of oxytocin. Furthermore, the researchers found that these rodents also have a larger number of receptors specific for oxytocin and vasopressin. As we mentioned above, vasopressin is found particularly in rats, where it exerts effects that resemble those of oxytocin.

Together with the neurobiologist Larry Young at the same university, Insel tackled the question which genes are responsible for the production of these receptors. During this research, they discovered an insertion in a gene of the monogamous prairie vole which is suspiciously absent in the polygamous montane vole.

To test whether this insert is linked to the difference in sexual behaviour, the researchers subjected the unfaithful montane vole to a kind of gene therapy. They incorporated the gene with the insert into the genomes of male montane voles. Indeed, after an injection of vasopressin, the genetically manipulated rodents suddenly established contact with their mating partners and even showed an interest in their offspring. Thus, with a single gene manipulation, the social behaviour of the animals could be turned around completely. How long until human females can apply this little improvement to their unfaithful husbands?

In contrast to oxytocin, there is relatively little research on the effects of the chemically related hormone vasopressin on human behaviour. However, most researchers in the field believe that vasopressin, too, is involved in our ability to establish bonds with other people.

One example is found in children with autism. This condition is characterised by a triad of impairments affecting communication, imagination, and social interaction. Scientists have shown that the level of vasopressin in people with autism is often extremely low.

CHAPTER 8

Dopamine – Casanova's Double-Edged "Secret Weapon"

He who renounces love entirely is no less ill
than he who desires it too much.
Euripides ($\sim$ 480–406 BC)

1 A HORMONE THAT MAKES US EUPHORIC

Having not quite enough of this substance can have grave effects on the brain and lead to Parkinson's disease or schizophrenia. The right dosis can get the emotions flowing. A little too much of it can turn you into a Casanova against your will. We are talking about another important hormone, namely dopamine.

What is so special about this hormone? As scientists have found out, there is plenty of dopamine in the blood of people in love, and it appears that the molecule – known to chemists as 2-(3,4-dihydroxyphenyl)ethylamine – plays a leading role in the games of love. It increases the physical motivation and on top of that it makes us euphoric.

As we know today, dopamine acts directly upon the primitive part of our brain, the limbic system (see Chapter 1). The more dopamine arrives there, the better we feel. Dopamine is therefore known as the "happiness hormone" as well. This is the positive side of the coin.

In recent years, some researchers have tried to investigate the effects of this hormone more closely. One of the pioneers among these hormone hunters is the Italian Donatella Marazziti, a psychiatrist at the University of Pisa. Together with her colleagues she runs a research project to examine the processes occurring in the bodies of people in love. A group of appropriately enamoured students served as the human guinea pigs. "Racing heartbeats and sweaty hands are only the superficial signs of being in love," she says. On the biochemical level, much more is happening.

In their research, Marazziti and her coworkers have focussed on the messenger chemicals in the blood. They found that these substances, which have a direct effect on the psyche, are found in larger amounts when people are in love than when they are not. The enamoured students showed particular large quantities of dopamine in their blood. This messenger makes people open up towards others to a larger extent than they would normally do. "We can conclude from this that the body of someone in love produces particularly large quantities of dopamine," Marazziti summarises.

1.1 Excess Amounts of Dopamine can Lead to Pathological Addiction to Love

If there is not enough dopamine in the blood, that loving feeling cannot develop properly. But an excess of the hormone can have negative effects, as the example of Casanova demonstrates. Marazziti says about her famous compatriote: "He must have had an absolute excess of dopamine." Presumably, this is the biochemical reason why his desire became compulsive. "In such cases we can call it an addiction, caused by too much dopamine," she diagnoses.

In contrast, people with the opposite problem need to take dopamine medication in order to harmonise their love balance. Marazziti believes that many people are unable to develop feelings of love simply because of their biochemical set-up. What makes the situation even worse is when the occipital areas of the brain, which are responsible for "sorting" emotions, cannot fulfil their task properly.

"In those cases, restlessness and nervousness, the typical side effects of being in love, can only be processed very slowly," she says. Administering therapeutic doses of dopamine could help such people.

The Canadian researcher Dennis Fiorine, based at the University of Vancouver, has investigated the biochemistry of dopamine using experiments with male rats. He applied a technique known as microdialysis, which involves implanting a small probe into the brain area of interest. Using very thin teflon tubing, which did not limit the movements of the animals, he was able to take samples of the brain fluid continuously and to analyse them rapidly.

During the experiment, the probe-carrying male rats were presented with a female rat in heat, which made the dopamine level of the males go up by 90 %. During the ensuing copulation, the level rose by another 10 %.

1.2 Male Rats Show a Pronounced "Coolidge Effect"

The most intriguing outcome of the studies, however, was a side effect which the researchers had not at all anticipated. Fiorini and his coworkers continued to measure the dopamine levels of the male rats after copulation. As expected, the levels decreased, despite of the presence of the female, dropping back to normal values. However, when the researchers presented a different female, the dopamine levels took off again, reviving the virility of the male rat.

This phenomenon is known in the scientific literature as the "Coolidge effect", named after the 30th president of the USA, Calvin Coolidge (1872–1933) (Figure 16), to whom the following anecdote is attributed: Mr. Coolidge and his wife visited a farm and were shown around in separate groups. Mrs. Coolidge noticed a cock copulating a hen and asked how frequently this happened. Being told that the cock did this up to twelve times a day, she reportedly said: "Tell that to the president." When the president reached the same spot and heard of those miraculous deeds, he asked: "Always the same hen?" Being told that it was a different one each time, he replied: "Tell that to Mrs. Coolidge."

According to Fiorini, several of these dopamine cycles could be produced with almost mechanical precision. In terms of evolutionary biology, the Coolidge effect represents a logical strategy to increase a male's reproductive success. As a male can hardly improve his chances of reproduction by repeated sex with the same female, there appears to

Figure 16   *Calvin Coolidge*

be a mechanism in the brain that deflates his libido after a while and diverts it to fresh opportunities.

Some researchers believe that the Coolidge effect is also responsible for the bedroom lull affecting many couples after a number of years. We don't subscribe to this interpretation. To us it appears more likely that psychological reasons, whose causes are found in the way we deal with each other, have a stronger influence. As the Austrian historian Erika Weinzierl noted: "Indifference towards the partner is the beginning of all evils."

CHAPTER 9

Serotonin – the Happy Messenger in the Blood

Love is as harmless as a spoon of hydrochloric acid on an empty stomach.
Charles Baudelaire (1821–1867)

1 A MOODY MESSENGER

As we briefly mentioned above, serotonin does not exactly count as a hormone, but as a messenger substance. However, as its effect on our moods strongly resembles the one of hormones, we would like to include it in our discussion. So what does this would-be hormone do, where is it produced, and how do we notice its effects or its deficiency?

Chemically speaking, serotonin is a biogenic amine. More precisely, it is 5-hydroxy-tryptamine, an indole derivative. Its biosynthesis, which occurs both in the central nervous system and in peripheral locations including the lungs, the spleen, and the lining of the guts, starts from the precursor L-tryptophane, which is one of the 20 amino acids that make up proteins. Among these, it is one of the few amino acids that we humans cannot produce in our metabolism, and which we have to take up with our food. Sources of serotonin or tryptophane include various kinds of fruit such as banana, pineapple, strawberries and raspberries. Sesame, rice pudding, and chocolate also provide us with serotonin.

Serotonin affects primarily the brain, where all our feelings arise as well. In an adult person, the body contains around 10 milligrams of the substance, which is the amount one needs for psychic stability. If the serotonin supply decreases below that level, the mood literally falls with it. The consequences include lack of drive, sleep disorder, anxiety and depression. Serotonin also influences our appetite and sensitivity to pain to a great extent. Long-term deprivation of serotonin is thought to cause serious illness, while acute lack of it can trigger some paradoxical reactions – including the "microparanoia" of people in love – as we will see below.

1.1 Sometimes Michael Craves Chocolate

Bianca has already noticed it: Michael has a weakness for chocolate. If it overpowers him, there won't be much left of the slab of chocolate he just started. "I suddenly had such a craving for it," he apologises each time. As it doesn't happen too often, Bianca responds to these "fits" with a smile.

Chocolate mostly consists of relatively simple components like fat, in the form of cocoa butter, cocoa powder, milk powder and sugar. These account for its high energy content – 100 grams of the dark delicacy harbour hundreds of kilocalories. Then there are additional flavourings such as vanilla, cinnamon and cloves. Finally, there are an estimated 600 further substances present in small amounts, including theobromin, caffein, and tryptophane – the precursor for our "happiness hormone" serotonin. There are also small amounts of 2-phenylethylamine, which will be the subject of the next chapter.

Apparently, chocolate increases the level of serotonin in the brain, thus boosting its ability to pass impulses from one nerve cell to the other. Quite possibly, there may be a deep reason why chocolate consumption peaks are in winter. The short days and lack of sunlight can make many people feel a low mood, known as "seasonal affective disorder" or SAD. It is thought that serotonin levels in the brain are involved in this phenomenon, hence the urge to chase away the winter blues by eating chocolate.

Richard Wurtman and his coworkers at the Massachusetts Institute of Technology (MIT) have recently found that a change of diet can cause stunning changes in the biochemical balance of the brain. They also discovered some connections between serotonin levels and light. For instance, they found that serotonin is involved in the control of sleep-wake rhythms. Further studies revealed that the concentration of serotonin in the brain depends on the duration and intensity with which our eyes are exposed to light. This finding links the compound to the biological clock mechanism, which not only controls the 24 hour cycle of animals and human beings alike, but also measures the changing day length and thereby controls seasonal adaptations, including hibernation in animals.

Thus, even though chocolate contains only small amounts of tryptophane, it remains a tasty and mild antidepressive agent, which can help many of us through the gloom of winter.

1.2 Pathological Love – Linked to Serotonin Deficiency

Does the following situation sound familiar to you? You have left your house to start a vacation. Just as you are about to get into the car, a thought haunts you. "Did I really switch off the coffeemaker, pull the plug of the TV, and close the window in the bathroom?" Thus, there is

Figure 17 *Donatella Marazziti*
© Donatella Marazziti

no way around it, you go back into the house and repeat your routine checks. Similarly, you may have checked the luggage several times, especially the passports, tickets, and other paperwork.

People who manically check everything before leaving may be suffering from a mild form of "control freakery." But what does that have to do with our topic?

"A lot," says Donatella Marazziti, the Italian psychiatrist whom we met earlier on (Figure 17). Over the last ten years, she has been studying people who compulsively repeat actions or thoughts. This includes, for instance, obsessive hand washing. In her studies, the researcher came across a much more common "disease."

1.3 Microparanoia – The Love Sickness

"People in love are also fixated on an object – the loved person," Marazziti explains. Occasionally, this mysterious state can be compared to a compulsive neurosis. She concluded that it must be possible to detect this state via the messenger serotonin.

In order to explore the effects of serotonin, the researchers took blood from a group of enamoured (female) students and compared their serotonin concentrations with those of patients suffering from compulsive neurosis. Surprisingly, the serotonin levels in both groups were comparable, both around 40% below the normal value.

Marazziti also investigated another group of female students who had left the excitement of the first phase of being in love behind, and she found out that the serotonin levels were approaching the normal values again. Being in love, she concludes, really does drive you a little bit crazy. She created a new name for this state of mind: microparanoia.

CHAPTER 10

Phenylethylamine – the Stuff that Makes the Soul Jubilate

Love is a pleasant state of partial inaccountability
Marcel Aymé (French writer, 1902–1967)

1 WELCOME TO THE ROLLERCOASTER OF EMOTIONS

On our round trip through the world of hormones we have now reached the last stop, phenylethylamine, or PEA for short. Like the other hormones, PEA is produced by our body, but it is also found in the oil of bitter almonds and – as already mentioned – in chocolate. Like serotonin, it has a strong psychic effect, which helps to explain why it is found as the core structure in a variety of hallucinogenic drugs.

A British study revealed the first indications of a strong psychogenic effect of PEA. It showed that depressive people often have a below average concentration of PEA in their blood. Conversely, subsequent studies showed that the concentration shot up by 77% after physical activity, which was linked to positive psychological effects.

In certain ways, PEA can be compared to adrenaline, as it increases blood pressure and pulse rate in similar ways. It turned out that the PEA values of skydivers were significantly higher just after jumping. Similarly, increased PEA levels have been reported after rollercoaster rides.

Apparently, the same is true for the "rollercoaster of emotions," as even mental sexual stimulation – be it from reading romantic fiction or from daydreaming – can provoke a rise in PEA levels.

1.1 The Elevator to Cloud 9 Smells of Fish

Some researchers in the US have even suggested that PEA might be the trigger for romantic love, which must sound surprising to chemists. Why should phenylethylamine, of all chemicals, an oily liquid smelling of fish and ammonia, act as the elevator to cloud 9 for people in love?

Other researchers, including the Australian chemist Peter Godfrey at the Monash University in Melbourne, gather that PEA is responsible for the sweaty hands, lumps in the throat, and butterflies in the stomach of people in love. "One day it might be possible to make synthetic drugs that conjure up the euphoria of first love," he speculates. However, he does not plan on getting involved in such projects.

"We're more interested in the potential of this hormone in the medical area," he emphasizes. Thus, there are indications that PEA might one day serve in the treatment of motor disorders such as Parkinson's disease. These investigations are far from being completed.

Furthermore, it remains unclear whether PEA plays its role alone or in a cascade of biochemical reactions which might involve both neurotransmitters and the cuddle hormone oxytocin. It is undisputed, however, that the wave of feelings starts from the brain and spreads throughout the body within fractions of a second. Thus the comparison with mood-enhancing drugs is obvious.

But no matter how intoxicating this self-made drug may be, it doesn't last long. After two to three years at the very latest, most researchers believe, the nerve ends in the brain will have adapted to the increased levels of PEA. The excitation fades, and the phase of being in love is over, at least from the point of view of neurochemistry. But that doesn't have to be a disadvantage.

"For some it is the end of love, and boredom begins – but for others it is only the beginning," suggests the Polish chemist and educationalist Janusz Wisniewski of Warsaw University. In his view, this phase represents the transition from romantic love to the complex happiness of a mature relationship.

The New York anthropologist Helen Fisher also voices her opinion on this sensitive issue. According to her, PEA plays an important role in evolutionary history. It holds a couple together, until their child has overcome the difficult first few years.

In archaic ethnic groups like the Australian Aborigines, the Inuit, and the Amazon Indians, children are typically born at four-year intervals. On the other hand, an investigation of 61 different cultures in today's world has shown that divorce rates peak close to the fourth year of marriage. Sadly, the anthropologist did not reveal in which ways chocolate could change these statistics.

CHAPTER 11

The Chemistry of Birth Control

Sex is the rhythm of the Universe,
the most enjoyable activity
that men and women engage in together
apart from reading, of course.
Curt Leviant (US writer)

1 HOW ABOUT THE PILL?

One could write entire books about birth control. In the context of our "Chemistry of love," however, the issue is a marginal one, which we will only cover very briefly. Therefore, this chapter does not claim to be complete with respect to the various practices of birth control, if only because some of them (*e.g.* calendar method, condoms) have only very little to do with chemistry (apart from the latex-making chemistry required for condom production).

Nevertheless, we will attempt a brief sketch of the history of birth control, which runs like a red thread through the cultural history of humankind. At the end we will arrive at the contraceptive pill, which has now been on the market for more than 40 years and which is a chemical product that we should mention not least because of its global importance and its phenomenal success story.

1.1 Birth Control – Almost a Never Ending Story

It has been reported that people already investigated methods of contraception in ancient Egypt about 4000 years ago, in order to get the better of nature. Instead of our birth control pill, the Egyptians used ground pomegranate seeds, which they combined with wax to roll vaginal suppositories. The principle of this method is indeed impressive, as we now know that pomegranates contain a plant-produced estrogen which, in theory, can prevent ovulation just like the modern pill. Sadly, there are no reports on the success rate (or otherwise) of this early method.

Another recipe from the same era, which the British archaeologist Sir William Flinders Petrie discovered during excavations south of Cairo in 1898, sounds rather revolting. It recommends quite seriously to introduce a mixture of honey, sour milk, and crocodile excrement (!) into the vagina in order to avoid pregnancy.

Ironically, even this recipe may have worked well, as the sticky honey would have blocked the sperms from accessing the uterus, while soured milk has some sperm-killing effect. However, the Egyptians would have been well advised to leave the crocodile excrements alone, as they are alkaline, not acidic, and thus create a positive environment for sperms. Not to mention the hygienic problem. This component arose probably from cult rather than science, as the Egyptians revered crocodiles, along with cats, as holy animals.

Moreover, around 1550 bC, the Egyptians came up with a recipe for a kind of tampon, supposed to avoid unwanted pregnancies for up to three years. It reads: "acacia tips, finely ground, are to be spread onto a fibre puff with dates and honey, then led deep into the womb." This instruction, too, reveals some real insights the early Egyptian physicians must have had. The sprouts of the acacia ferment and release lactic acid, which is also used in today's contraceptive gels.

1.2   Just Crouch Down and have a Good Sneeze

Hippocrates (460–377 bC), by far the most famous physician of the ancient world, writes in his opus on the nature of women: "If a woman does not want to conceive, she should make it a habit after each intercourse to let the semen fall out." Sadly, he did not leave any more specific instructions. Soranus of Ephesus, who worked as a physician in Rome at the beginning of the second century AD, reveals a trick: "The woman should hold her breath in the moment when the man ejects the semen, and pull back. Then she should get up, crouch down and sneeze strongly." Sneezing powder anyone?

Similar advice came from the Persian physician Rhases, who lived around the year 900 AD. He advised women who wanted to avoid getting pregnant, "to push the navel strongly with the thumb, in order to eject the semen from the vagina." Such concepts appear to have survived across the centuries, as it is reported that even in 19th century Austria women tried to rid themselves of the sperms received by performing snake-like movements.

In the Far East, too, people have been thinking about how to avoid unwanted pregnancies from very early on. Chinese sources from the 7th century AD recommend to heat mercury in oil for a day, then swallow a droplet of the concoction in the morning on an empty stomach. The Teutons are said to have used extracts of ivy, yarrow, plantain, white poplar, and pimpernels, while the Shoshone Indians in Nevada applied the root extract of a local desert plant as a contraceptive.

But only since around 1830 women have been able to consult their doctors and hope for sound advice. One of the pioneers of birth control was the American physician Charles Knowlton (1801–1850). In his book entitled "The fruits of philosophy" he recommended to wash the vagina with cold water or alum, vinegar or a solution of zinc sulphate immediately after intercourse. Like the ancient Egyptians before him, he counted on the spermicidic effect of acidic environments. In France, the vinegar wash is said to have been well known and widely used in Knowlton's time.

The modern era of industrially produced and commercially distributed contraceptive products only began some 120 years ago, starting from England. The blockbusters of the time included dissolvable chinine suppositories made of chinine and cocoa butter, which are reported to have been available at two shillings a dozen. Although their effect was rather unreliable, these so-called "Rendell's" (after the London chemist W.J. Rendell) were very popular up until WW II.

Only from the early 1950s onwards did the chemical industry bring effective spermicidal preparations to the market, improved versions of which are being used as suppositories, cremes, and foams to this day.

1.3 A Big Leap for the Biology of Small Eggs

Quite apart from these developments, the early 20th century also witnessed a gold rush of new discoveries in hormone research. At the beginning of the 1930s, the German chemist Adolf Butenandt (1903–1995) elucidated the chemical structures of the female sex hormones, for which he was awarded the Nobel prize in 1939 (Figure 18). Soon after that, progesterone was introduced to the treatment of women with previous miscarriages, or with serious menstruation problems.

Even then, it was already known that the hormone acts as a contraceptive. However, it could not be used as such on a large scale, because it tends to get digested very rapidly if taken orally. It was only in 1951 that the chemist Carl Djerassi, together with the biologist Gregory Pincus achieved the major breakthrough for the development of modern contraceptives.

Figure 18   *Adolf Butenandt*
               ©GDCh.

Developing an artificial hormone called "norethindrone," a derivative of progesterone, Djerassi had hit two birds with one stone. As it turned out, norethindrone is much more reliable than the natural progesterone if taken orally. It effectively stops ovulation from happening, but on the other hand the estrogens in the pill make sure that menstruation still occurs on time. Thus, Djerassi succeeded in mimicking the natural female cycle (see Box on p. 91).

Today, the pill is without doubt the most widely used contraceptive in the world. Many millions of women take it every day. Its reliability is measured with the so-called Pearl index, which is defined as the number of pregnancies found in 100 women who have been using a given method of contraception for one year. The Pearl index for the birth control pill is between 0.2 and 0.5, depending on how carefully women stick to the instruction to take the pill exactly at the same time every day.

Today's pills contain the hormones estrogen and gestagen in different combinations and doses. Some alternative methods such as vaginal rings and contraceptive patches also contain hormones. In contrast, the modern mini-pill only contains a very small amount of gestagen. Therefore, it does not inhibit ovulation. Instead, its effect is based on the changes to the lining of the uterus, and the cervix. This stops the

Figure 19   *Carl Djerassi*
©Michigan State University/Carl Djerassi.

sperms from reaching the uterus. Even if an egg did get fertilised, it would not be able to implant itself, as the lining of the uterus would not be prepared for this event.

CARL DJERASSI – "FATHER OF THE PILL"

The year 1951 marks the breakthrough for the development of the first reliable oral contraceptive. In this year, the Austro-American chemist Carl Djerassi (Figure 19) applied for a patent for an orally active derivative of the female sex hormone progesteron to be used as a contraceptive. Djerassi set out to synthesize cortisone, a hormone of the adrenal cortex, and by chance invented the hormonal contraceptive, which was to become one of the most used drugs in the world within the next decades. Since then, Djerassi is known as the "father of the pill."

Born in Vienna, he spent his early childhood in Sofia (Bulgaria), then his school years in Vienna again. In 1938, he emigrated to the US, where he studied at several universities. At the age of 22 he obtained his PhD at the University of Wisconsin at Michigan. His research was focussed on sterols, complex alcohols such as choles-terol, which are found in all animals and plants.

As a group leader with the company Syntex, Djerassi achieved the first synthesis of a "steroidal" oral contraceptive on October 15, 1951. It was a surprise victory for an outsider in a race that involved many of the major pharmaceutical companies. The first studies of the drug were carried out by animal testing and pharmacological screening. Then there were large scale clinical studies conducted in Thailand,

Puerto Rico, Mexico, and the USA, some of them under the patronage of the World Health Organisation. Many women, especially from rural populations, volunteered to take part.

In 1957, the pill was licenced in the first federal states of the US. From 1960, it was available nationwide. It was introduced in many other countries including Britain in the following year. After some teething problems, it soon rose to become the most popular contraceptive of all.

The rise of its popularity was neatly mirrored in the drop in birth rates throughout the industrialised world. The baby boom of the post-war era was followed by a steep fall from 1964 onwards. (In Britain, birth rates dropped by 35% following the 1964 peak.) This phenomenon is all the more remarkable if one takes into consideration that the late sixties were a time of general sexual liberation.

1.4   Why is There No Pill for the Male?

Over the last 40 years, there have been repeated attempts to develop an effective oral contraceptive for men as well. In spite of many efforts, nobody has succeeded in bringing such a product to the market. In the 1960s and 70s, experiments were carried out with cotton-seed oil. Although it did achieve a contraceptive effect, this preparation ultimately failed, because it led to permanent infertility in around 30% of the cases.

A major factor is the psychological aspect of the problem, relating to the role of men. Back then, the willingness of men to look after contraception wasn't very pronounced, as these issues were perceived as a female domain. And ultimately, if the man failed to ensure his contraception was effective, the woman would be left – quite literally – holding the baby. As attitudes have changed over the last years, observers think that a male pill could have a chance in today's market.

Some of the more promising approaches being investigated now are based on gestagens. There are clinical trials with volunteers who take gestagen pills. These have an effect on the pituitary gland, suppressing the release of another hormone which in turn is necessary for the production of sperms.

The downside is that the pill also suppresses the production of testosterone. As we have seen in chapter 6, this hormone is essential for the male libido and many other things, so the volunteers have to get their testosterone level topped up by intramuscular injection once a month. Testosterone taken orally would be degraded in the liver. The current trials are aimed at establishing the preparation is well tolerated and effective. However, it may take years before a male pill might reach the market.

CHAPTER 12

Menopause: When the Hormone Supply Falters

To love a person means to agree to grow old with them.
Albert Camus (French writer, 1913–1960)

1 WILL YOU ALWAYS LOVE ME?

For Bianca (26) and Michael (30) there is no reason to worry (yet). We hope that the two of them have a number of carefree years ahead of them, and that they will still love each other when they reach the point where their bodies don't always comply with their wishes. Their friends know about the true reasons of their love, which is quite obviously not only built on the quicksands of physical desire.

"Will you always love me?" Bianca once asked Michael this questions immediately after an amorous encounter.

"Yes, why do you ask?" he replied, astonished. She bent over to him. "Because I'm just thinking ... that one day we will be old."

Michael thinks for a moment: "You mean, when we can no longer ... so often ...?" She puts her finger on his lips. "Exactly."

Michael hugs her. "Well, we can always have a cuddle."

This brings us back to the chemistry of love. When we consider the complex interplay between well-being, experience, and hormones, we cannot avoid the much talked about (especially among women) topic of the menopause.

There are some women who already suffer from the specific menopause symptoms at the age of 40, while most women first experience them between the 48th and the 52nd year. The typical phenomena include:

- headaches,
- sleeping and circulation problems,
- vertigo,

- sweating,
- bouts of depression
- racing heartbeats,
- heat flushes,
- dryness of the vagina.

The root cause of these symptoms is the continuous decline of the estrogen level in the blood beginning around the 40th year. Thus, the menopause is a hormonal mirror image of puberty. As a consequence of the declining production of the sex hormone estrogen, ovulation becomes rarer and eventually the function of the ovaries ceases entirely. Menstruation will become irregular and eventually fail to appear at all.

The symptoms connected to the hormonal change begin slowly at first, sometimes in bursts, until they become manifest. The uterus and the mucous membranes of vagina, urethra, and bladder recede.

To many women, the most bothersome symptom is the appearance of creases, especially around the eye, reflecting the declining ability of the skin tissue to store humidity. The slackening of the breasts is also experienced as a loss. Furthermore, the bones begin to become brittle, as there is no longer enough estrogen to keep the metabolism of the bones going. Therefore, the risk to suffer from osteoporosis increases drastically from menopause onwards. Another important symptom is the reduced secretion inside the vagina. As this can lead to pain during intercourse, many women lose interest in sex at this point.

1.1   Is Hormone Replacement Therapy the Way Out?

The most efficient treatment that can alleviate all the ailments connected to the menopause is hormone replacement therapy (HRT). It is undisputed that hormonal treatment not only addresses the symptoms but can also minimise the reduction effects in the sexual organs and the risk of suffering from osteoporosis.

On the other hand, medical studies have shown that HRT slightly increases the risks of thrombosis and breast cancer. According to the results available so far, one woman out of 100 who take hormones over more than five years will experience one of these serious side effects. On the plus side, the studies also showed that the risks of bowel cancer and osteoporosis are slightly reduced. The same may be true for Alzheimer's and Parkinson's disease. Thus, in purely medical terms, benefits and risks of HRT are approximately balanced out.

Apart from the medical studies, there is the individual perception of each woman. Most say that they feel significantly better with HRT. This clearly includes the chemistry of love, as the hormone therapy also improves the fluid secretion in the vagina. Estrogens will also improve the libido, and most women will feel younger.

Should a woman take hormones during menopause or shouldn't she? This is a thorny question, which a doctor can never answer with a simple yes or no. Much less would we dare to issue any recommendations. Every woman will have to make this decision for herself. We would suggest, however, that a woman should not simply collect a prescription, but that she should have a thorough discussion with her doctor, in order to establish a profile of her individual risks. If for example, there have been cases of heart disease, thrombosis, lung embolism, or breast cancer in the family, caution and regular check-ups are advised.

There are alternatives to the conventional HRT. One of them, the drug "tibolone" (also known as "livial") does not only contain estrogen, but also progesterone, plus very weak doses of male hormones. There are good reasons for developing such a cocktail of hormones. In the past, the male hormones (the androgens), which are also produced in the ovaries, were often neglected. "Today we are a bit wiser and we know that male hormones are very important for the well-being of women," says Christian Egarter, a gynaecologist at Vienna, Austria. A deficiency of these hormones can affect the libido and reduce mental ability.

As with conventional HRT, the estrogens in tibolone will help women to avoid hot flushes, sweating, and osteoporosis. In contrast to conventional HRT, tibolone does not lead to bleeding, a difference which many women appreciate. Recent research has also shown that nearly all women treated with tibolone experienced a significant increase in libido.

It also appears to have a beneficial effect on heart and circulation, according to research presented at a conference in 1998. The researchers had fed laboratory animals with an extremely cholesterol-rich food in amounts which would normally lead to arteriosclerotic changes in the blood vessels after a very short time. If they added tibolone to the food, the changes failed to appear, although they still appeared when only estrogen was added. Moreover, recent studies have suggested that tibolone does not promote breast cancer. Both animal experiments and studies with cell cultures derived from human breast tumours suggest that the drug even slows down or suppresses tumour growth.

1.2 Drugs from Nature's Treasure Trove

Many women place their hopes on plant preparations. While we would not want to dismiss this kind of treatment, these alternative drugs can only alleviate the symptoms in the initial phase of the menopause.

Figure 20   *Black cohosh (Cimifuga racemosa)*

Ultimately, they fail to address the underlying problem, namely the fact that the hormone sources will slowly run dry.

Thus, drugs derived from the black cohosh (*Cimifuga racemosa*) (Figure 20) are mainly effective at the onset of menopause, when hot flushes, sweating and vertigo are the dominant problems. St. John's wort has a mood-improving effect, while tranquilising drugs such as valerian can improve sleep. The sweating can be treated with sage.

Some women rely on the natural effects of soy. This alternative remedy is based on the observation that women in Asia only rarely complain about the ailments related to menopause. (However, this might be a cultural phenomenon – apart from exceptional cases, they don't complain about their husbands either!) Researchers have now shown that soybeans, which are a staple food in most Asian countries, have estrogen-like effects. Thus, if anybody wants to replace their hormones by including soy-based food in their menu, there is nothing wrong with that. However, taking concentrated soy extracts may carry risks that are as yet unexplored.

1.3   Not only Women

In contrast to women, who normally talk openly about the ailments of menopause, the topic still remains taboo among men. Obviously, symptoms like declining virility, buzzing in the ears, vertigo, memory and concentration weakness, as well as mood swings undermine a man's confidence. Thus, men tend to keep quiet about or even ignore these natural side effects of age. It is often claimed that there is no male equivalent to menopause, but insights from medical research clearly contradict this claim.

However, it would be slightly misleading to speak of a "male menopause," as the hormone level doesn't drop nearly as fast in men as it does in women. While the changes in women are normally limited to the time between the mid-40s and the late 50s, the male hormone level declines little by little over a time of three to four decades.

"ANTENATAL EXERCISES" HELP FIGHT ERECTION PROBLEMS

There is a wealth of more or less useful advice on natural ways to fight erection problems in the literature, ranging from sports to changes of lifestyle. In the latter category, there are recommendations to replace fatty foods with a menu rich in bulkage and vitamins, and to reduce the intake of nicotine and alcohol.

A very different suggestion comes from the British medical researcher Grace Dorey from the University of Bristol. She encourages men to do a kind of birth preparation gymnastics. "Training of the pelvic floor muscles could help men with erection problems just as efficiently as viagra," she claims after completing a study involving 55 men who had suffered from erection problems for at least six months, and who agreed to repeat the exercises on a daily basis.

Within half a year, the problems disappeared in 40% of the participants, and an additional third of the patients reported significant improvements. Only the remaining quarter saw no success resulting from the treatment. Dorey also reports that any problems of dripping after urinating were improved by the gymnastics, and therefore suggests to recommend this kind of exercise not only to pregnant women.

Studies with a larger number of men will reveal whether her promise – that the moves are just as efficient as viagra – can be kept.

While many doctors recommend HRT for women, a similar treatment of men using the hormone testosterone is more controversial. Among those men with erection problems, only 10% can blame a reduced testosterone level for the disorder (see the Box above for an alternative cure). Therefore, the lack of the hormone must be established by chemical analysis, before a hormone treatment can be prescribed. A medical journal recently recommended to substitute testosterone only when the concentration in the blood is below 10 nanomol per litre and symptoms like anaemia, reduced bone density in the lower spine, and a shrinking of the prostate are observed.

A hormone that is more widely used than testosterone is DHEA (Dehydroepiandrosterone), a hormone of the adrenal cortex, which is regarded as an anti-ageing drug especially in the USA. The fact of the matter is that an average 60-year-old man only produces a third of the amount of DHEA that a 20-year-old produces. That's enough reason for millions of Americans to try and rejuvenate themselves with this hormone. But does science support this application?

Recent studies have shown that men with a DHEA deficiency and virility problems can experience a significant improvement of their quality of life and ability to perform when they take the hormone. Furthermore, although the manifold effects of DHEA in the body remain to be explored, several long-term studies have shown that men with reduced levels of DHEA have significantly higher mortality rates from heart attack and stroke than men with normal levels. In men with profound depression, DHEA leads to a marked improvement of their general well-being.

In general, however, the advantages of hormone therapy in men must be carefully evaluated. First of all, the possibility of an undiagnosed prostate cancer has to be ruled out. During hormone therapy, a close supervision by a doctor is necessary, in order to recognise any changes – especially in the prostate region – in time. This is not because the hormones might cause cancer – in fact they don't. But there are certain kinds of cancer cells which "feed on" hormones like testosterone and DHEA, a phenomenon which can speed up the growth of very small or nascent tumours. Thus, we are confronted with a dilemma similar to that of female estrogen therapy, which is linked to an increased risk of breast cancer.

The situation is similar for treatments with the growth hormone somatotropin. At the moment there are no reliable clinical results concerning the positive or negative effects of such a therapy. It appears to be justified only if the level of the hormone in the blood is found to be significantly decreased, and a decline of muscle power and physical performance is observed.

Now and again, there are reports of gynaecologists who recommend a pure estrogen therapy for men. Urologist Hartmut Porst from Hamburg, Germany, advises strongly against such a measure, as there is no scientific evidence for its benefits. He says that the loss of estrogen in men is always coupled with the decline in the testosterone levels. Thus, testosterone therapy, possibly in combination with DHEA, would be the first choice.

A healthy and balanced lifestyle contributes decisively to the preservation of a man's (and woman's) fitness. If you keep mind and body going, are more on the gourmet than on the gourmand side, appreciate a good drink, but only within limits, don't avoid a critical confrontation with the bathroom scales, and refrain from the use of vessel constricting drugs like nicotine, you're in with a good chance.

CHAPTER 13

Endogenous Opiates – the Chemistry of Euphoria

"We can only be cured of a passion
if we enjoy it to the full."
Marcel Proust (1871–1922)

1 IT'S NOT ONLY THE ADRENALINE THAT GETS MICHAEL AND BIANCA STARTED

When Bianca and Michael met at the airport after a long time apart, when they hugged and kissed each other, their hearts were racing like mad. One of the reasons for this side effect is the increased adrenaline level – a topic that we have already discussed above (Chapter 5).

US scientists have called the situation surrounding such kissing and loving experiences a "positive stress," and they were surprised to find that an intensive kiss not only involves lips and tongue, but also 34 other facial muscles. The whole process makes the face glow, while the skin is tightened to produce a younger look. At the same time the pulse rate increases from the normal 70–80 to 120 beats per minute, the peripheral vessels carry more blood, and a healthy blush appears on the cheeks.

However, they also found that a passionate kiss can get our body chemistry started in a way that turns the blood stream into a "high"-way carrying an incredible mixture of the body's own drugs. We will now have a closer look at this biochemical brew.

Surprisingly, it appears that the multitude of hormones and messengers that we have described above and which seemed to act like the ingredients of a perfect love cocktail, aren't quite sufficient to account for the chemistry of love.

100

1.1 Is Our Brain a Poppy Flower?

Even though you may be thinking that the potpourri of different messengers floating around our body is already diverse enough, we have not quite reached the bottom of the barrel yet. The one important group that remains to be discussed is known as the endorphins (from endogenous and morphine), a highly unusual group of peptides.

"Our brain is a shining white poppy flower, releasing opium to alleviate our pains," as the French Neurophysiologist Jean-Didier Vincent writes in his book "The Biology of Emotions." Believe it or not, every human being is a producer and consumer of drugs at the same time, and we get them legally and free of charge. People who take part in extreme sports such as marathon runs are almost turned into junkies. That may sound crazy, but it is not even an exaggeration.

We know about these incredible facts since 1974, when the US scientists Solomon Snyder and Candace Pert, working at the Johns Hopkins University in Baltimore, Maryland, made an unusual discovery, which of course created some lively debates (Figure 21).

The two of them had focussed their research on the properties of morphine. They wanted to find out how the drug exerts its effect. As "opium," the substance has accompanied humankind for around 6,000 years, as the ancient cultures already used the concentrated juice of the poppy capsules as a remedy against pain, depression, and a number of other ailments.

Figure 21 *Candace Pert (left) and Solomon Snyder ©Candace Pert and ©Solomon Snyder*

Even in World War I, soldiers received morphine in order to be able to forget their pain and anxiety. Today we know that the feeling of pleasant relaxation which the poppy juice provides will have to be paid dearly with physical and mental dependence and the subsequent decay of body and mind.

All the more spectacular was the finding that our own brain is able to release opium-like substances, namely the endorphins. Almost at the same time as other research teams, Snyder and Pert found out that the human nervous system displays specific receptors, into which the opiates fit like a key into a lock, and through which their effects are passed on immediately. With this, the mechanism of morphine action was elucidated in principle, and the main question that the researchers had pursued was answered.

However, the two were far from content with their achievement. As it often happens in science, the answer they found had thrown up a further, even more interesting question: Why on Earth does the human body have specific receptors for opiates? Has the drug use over millenia resulted in an evolutionary response, creating docking sites for them?

That can't be right, Snyder and Pert must have thought. They came up with a more far-reaching conclusion. If there are specific opiate receptors in the body, they hypothesised, then there must be endogenous substances to fit these receptors. But what might they be?

1.2   Researchers Go the Whole Hog

One year later, Hans Kosterlitz, the director of the pharmacological institute in Aberdeen, Scotland, and his student, John Hughes answered this question. The two of them had to collect more than 2000 pig brains from local slaughterhouses, in order to extract from them a tiny amount of a substance that had so far been unknown and that had a similar effect as morphine. They called the substance enkephalin, which politely refers to its location in the head, but doesn't mention the multitude of pigs involved in the research.

The discovery caused quite a sensation. Previous speculations about how cultural history might have brought the receptors into existence suddenly became irrelevant. Over the next few years, it even turned out that these receptors represent an ancient system (similarly to the one for oxytocin). It is found in the brains of all vertebrates and must have developed well before plants and animals even colonised dry land. Thus it clearly predates the appearance of the poppy flower, whose reputation was spectacularly cleared.

"At first glance, the brain morphines don't appear to have any resemblance with the morphine derived from plants," the US biologist Robert Ornstein comments the discovery. Endorphins are classified as neurohormones and chemically speaking they are polypeptides from the group of endogenous opiates.

The same is true for the enkephalins, even if they have shorter chains than most endorphins. As nerve messengers or neurotransmitters they have a direct effect on the perception of pain, by locking into the opiate receptors. This way, they can even completely block the transmission of pain for a certain time. The morphine molecule, in contrast, is a small molecule that has nothing to do with peptides (see Box on page 104).

1.3 Long Time No See?

There is a true anecdote from the experience of one of us (R.F.) that may help to illustrate the science of endorphins which we will continue to discuss below.

Sometime in the late 1990s I visited a fair at Leipzig. While I was looking at one of the cars on display, a man purposefully steered towards me from a stall located opposite.

"Hey, Rainer, old friend, how are you doing these days? Long time, no see!"

"I'm sorry but I don't know ..." I replied, rather baffled.

"Come on. Is your memory that bad? I'll give you a clue. VEB so and so at Dresden, where we worked together for a few years before the reunification. Yes it's me, your old colleague Horst.," the guy insisted in his unmistakable dialect from Saxony.

There was no other way out, I had to explain to the poor man that my first name was Rolf, not Rainer, and that I was from Western Germany and never worked in the GDR. The man looked at me wide-eyed and concluded that I must have a perfect *Doppelgänger*.

I can't judge whether the man I talked to just had a poor memory for faces, or whether I do have a real *Doppelgänger*. If the latter is true, I admit I would like to meet my long lost twin. It appears, however, that there is at least one other person who looks like me – even though we are not genetically related – to such an extent that other people can take us for one and the same individual.

Basically the opiate receptors in our brain are in the same tricky situation as the man I met at Leipzig. It is only because of a superficial similarity that they take two completely different molecules for one and the same. "In its three-dimensional shape, one end of the morphine molecule is indeed surprisingly similar to one end of the enkephalin molecule," Ornstein explains. Of course the opiate receptors in the brain were never meant to bind opiates. Strictly speaking, they are enkephalin receptors, whose natural "keys" are the endorphins of the brain (Figure 22).

COMPARISON BETWEEN ENDOMORPHINES AND MORPHINE

Endomorphines are a group of endogenous substances which share the pain-alleviating properties of the plant-derived morphines, have certain structural properties in common, and are produced by the central nervous system. Depending on the core of their molecular structure, scientists distinguish between three categories:

- Derivatives of pro-opiomelanocortin. Cleavage of this large precursor protein in certain cells of the pituitary gland and the hypothalamus produces beta-endorphin, a peptide of 31 amino acids.
- Derivatives of pro-enkephalin A. The best-known examples are leu- and met-enkephalin, small peptides with five amino acids each. These peptides are found in large areas of the brain and in the spinal cord, where they act as neurotransmitters.
- Derivatives of pro-enkephalin B. Mainly dymorphine and beta-neo-endorphin. They occur in various structures of the brain, but not in neurones which contain derivatives of pro-enkephalin A.

All these endomorphines are released by neurons. They exert their effects on targets in the brain or in peripheral sites via specific opiate receptors, which come in three classes, labelled with the prefixes χ, μ, and δ.

The classification of the receptors relies on their binding affinities for certain opiates and the morphine antagonist naloxone, as well as on certain pharmacological effects observed under defined experimental conditions.

The similarity between endomorphines and morphine arises exclusively from the overall three-dimensional shape, which can be created with completely different chemical architectures. While morphine is a relatively small organic molecule based on a mesh of four carbon-based rings, endomorphines are polypeptides, made of amino acid building blocks.

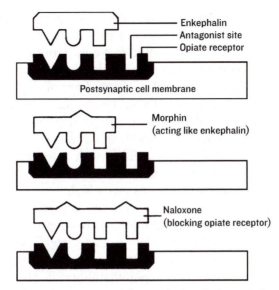

Figure 22 *Molecular interactions according to the key–lock principle*

"It is pure coincidence that morphine has a shape that happens to fit into the opiate receptor," Ornstein explains. The same holds for the synthetic drugs which trick the receptors in the brain with a set of obviously forged, but very efficient keys.

The opiate receptors are found in the entire nervous system. They are most concentrated in the spinal cord and in certain areas of the brain. However, they also occur in nerve endings, in the lungs, the bladder, and the guts. There is a conspicuous concentration of opiate receptors in the limbic system, the part of the brain that is responsible for moods and feelings. This explains why endorphins not only alleviate pain but also lighten the mood. Investigations have shown that people with depression often have a lack of endorphins.

2 ENDORPHINS – AN EMERGENCY CENTRE OF THE BODY?

The key question is: why did nature create these opiate mimics? Obviously, they help us to cope with stress situations of all kinds. Furthermore, they are able to trigger happiness, joy and even euphoria. The endorphin content in the blood of a woman giving birth, for example, is around ten times higher than normally. Thus, the body's chemistry finds a way to help mother and baby to cope with the stress and pain involved in the birth process.

In similar ways, endorphins also help us in other extreme situations. In an emergency situation, we may not even notice small wounds, which under normal circumstances would cause us pain. Another well-known phenomenon caused by endorphins is the peculiar happiness of marathon runners. One might call it a natural kind of doping.

However it's not just the extreme sports like bungee-jumping, paragliding or marathon which can help you get high on endorphins. People who keep fit and compensate for their sedentary office life by jogging regularly are rewarded for their effort with a feeling of lightness and freedom from all worries. An increased release of endorphins provides them with what is known as a "runner's high."

On the other hand, there are reports of joggers who suffer from withdrawal symptoms if an illness or other reasons keep them from running. More worrying still, the "runner's high" might mask strong pain, so that in extreme cases even symptoms of an imminent heart attack caused by the exertion might go unnoticed. This may have sealed the fate of the Greek messenger who ran from Marathon to Athens to report the victory of the Greeks over the Persians some 2500 years ago. Immediately after his arrival he collapsed and died.

No matter whether it's jogging, walking, cycling or swimming: all sports depending on endurance can rev up the production of endorphins. Scientific studies have shown that people suffering from chronic fatigue, depression, anxiety, or panic attacks benefit from these kinds of sport more than from drugs. The only requirement: several times a week they have to exhaust themselves for 30 minutes at least. Often, such a program results in a significant improvement of mood, which is apparently coupled to the improved physical and mental performance.

Is there a way to do it without exhaustion? Whoever hoped for a big "YES" from us will be disappointed. However, those who are content with more modest results, can get a small "yes" as a consolation prize. Indeed, there are far more gentle methods, including meditation, autogenous training, yoga, and even fasting, which ensure an increased production of various endogenous drugs in the body.

Even the increasingly widespread problem of anorexia has been linked to the dependence on endogenous opiates. Fasting provides the person concerned with a release of endorphins, which suppresses their depressions and anxieties. As endorphins can lead to addiction, patients continue to fast, even if they are close to starvation and have lost most of their strength. Apart from a psychological dealing with the individual and family problems, efficient therapy also requires a rehabilitation program, just like any drug addict.

2.1 Mice Turn into Little Cowards

A while ago, the scientific journal "Proceedings of the National Academy of Sciences of the USA" published an intriguing report by Donald Wells Pfaff and his colleagues at the Rockefeller University, New York. By means of genetic modification these researchers had rendered mice incapable of producing endogenous opiates, thus breeding rodents lacking endorphins.

The results were dramatic. The smallest fright made the animals freeze. In large rooms they stayed close to the walls and showed other typical signs of fear. The researchers conclude that endogenous opiates in humans as in mice might have the purpose of suppressing fear and anxiety.

2.2 Faith is the Best Medicine – Endorphins and The Placebo Effect

Even though it has been known for centuries and is being used in medical studies every day, the placebo effect was regarded as a myth until very recently. The mysterious phenomenon that a simple sugar pill with no drug content whatsoever can induce measurable improvement of a patient's well-being posed a conundrum to generations of medics. It appears that, when triggered by a placebo treatment, our body is able to produce its own drugs that efficiently induce the healing process of the disease in question.

The American physicist Howard Brody, a professor emeritus of the University of Pennsylvania, has investigated and described the placebo effect in countless studies. He thinks that there are three keys which "release the body's inner pharmacy":

- Expectation: physical changes happen when we anticipate them mentally.
- Conditioning: experiences from the past create a behavioural pattern that exerts its effect in the present.
- Significance: the way in which we interpret a disease influences the course it takes.

Equally crucial is the verbal "packaging" presented by the doctor. Statements like "These pills are extremely effective. You will soon notice that the pain ceases," spoken with the true ring of conviction enhance the expectation of the patient and thus the placebo effect. By contrast, phrases like "Try that, maybe it helps …" are more likely to act as placebo-antagonists.

A key requirement for the placebo effect is, of course, that the patient expects the drug, *e.g.* an apparent analgetic, to relieve his ailment. This anticipation arises in the limbic system, where the endorphin system is located as well. Might endorphins be involved in the placebo effect? A first piece of evidence for this link is the finding that the effect can be blocked by the opiate antagonist naloxone, which we will soon discuss in more detail.

Moreover, placebo effects appear to target the brain directly. Thus, functional MRI images taken after application of fake drugs revealed changes of the brain metabolism that were similar to those observed after administration of real analgetics.

IS ACUPUNCTURE BASED ON THE PLACEBO EFFECT?

Acupuncture is a part of traditional Chinese medicine. It is based on the concept that targeted excitation of certain spots can achieve a therapeutic effect.

When China began to open up at the end of the Cultural Revolution, groups of Western physicians visited the country in order to study the use of acupuncture as an anaesthetic in surgery. They remained mystified with respect to how this approach might work. However, they found that the use of the needles can reduce the amount of anaesthetic by around 20%. However, it was also found that only six out of ten patients respond to this treatment at all.

Researchers set out to find the mechanisms responsible for the pain alleviating effect. In 1976, they found that endogenous opiates obviously play a crucial role in that. Further research showed that acupuncture leads to a measurable relaxation of certain muscles and increases the blood flow.

Today, acupuncture has gained a foothold in Western countries as an alternative way to treat pain. It is mainly used to treat migraine, neuralgias, headaches, pain of the back, the face and the movement apparatus, and vegetative disorders. It is also used successfully to treat disorders of breathing and digestion.

---

3   ENDORPHINS IN PAIN RESEARCH

During the 1990s, researchers found out that the immune system and the nervous system cooperate in the origin and control of inflammatory pain directly at the location of its source. The coordinated response also involves immune cells migrating into the damaged tissue. It is also thought that endogenous opiates play an important role in the process.

The endorphins in inflammation are produced by the immune cells and bind to opiate receptors on the nerve fibres, which pass on the pain signals, thus reducing the sensitivity to pain during the course of the inflammation reaction.

For medical practitioners, the finding that opiates are also active at the periphery, not only in the central nervous system, opens up completely new opportunities for therapy, including the use of strong painkillers directly at the location where the pain originates. This way, the undesirable side effects of such drugs on the central nervous system can be avoided. Numerous clinical studies have shown that the approach works.

After endoscopic operations of the knee-joint, for instance, the injection of morphine directly into the joint achieves a significant alleviation of the pain almost without side effects. Clinical studies show that this treatment reduces both the intensity of the pain and the use of further painkillers after the operation.

Another clinical study involved 45 women who had both fallopian tubes interrupted by endoscopic operation for the purpose of sterilisation. After the surgery, the doctors injected an opioid (an artificial substance with morphine-like effects) into one of the operated areas, and a placebo (salt solution) into the other one. Neither doctors nor patients knew which side was which. The patients reported significantly less pain on the side which was treated with opioids. "As minimally invasive laparoscopic operations are increasingly popular, such results are very important," says Michael Schäfer from the Free University of Berlin.

Recent research from Schäfer's group shows that local application of opioids against chronic inflammatory pain, *e.g.* in arthritis not only alleviates the pain, but also reduces the inflammation. "The application of opioids directly into the inflamed joint could become a new alternative to the standard therapy which acts via the central nervous system," says Schäfer. However, he warns that repeated injections into the joint cannot be carried out, as they are linked to risks including bleeding and infection.

Therefore, he hopes for the development of opioids that are active only in the periphery but not in the central nervous system, as such drugs could be administered via the blood stream or orally. Pharmaceutical companies have already started with the development of such opioids, but the first trials have brought contradictory results. "This may be because the first generation of compounds still needs to be improved," says Schäfer.

3.1 Opiates Linked to Near Death Experience?

It is now a widely held belief that endogenous opiates, *i.e.* the endorphins and enkephalins, are involved in the phenomenon known as "near death experience." In many reported cases of such "dying experiences," people have the impression that their soul is getting separated from their physical body and is hovering above the scene. Thus they are looking at their own body from above. For example, a hospital patient might witness as a hovering outside observer how doctors and nurses are fighting for his or her life.

Only moments later, a kind of tunnel appears to open. The patient often feels himself drawn inside, floating towards a bright but not blinding light at the end. These are the most frequently reported experiences of people who have been close to death. Furthermore, there may be a so-called life review before the tunnel effect. The patient may see important scenes of his life passing by as in a movie, where he is only an uninvolved observer, looking at himself from a distance.

Some scientists consider it quite natural that a euphoric state of mind can be achieved not only in extreme sports such as triathlon, but also in near-fatal situations such as freezing or drowning. While this sounds plausible, it remains unclear why the experiences of many different people are so similar. And it will certainly be difficult to find conclusive evidence to show that endorphins are involved in this. After all, when a patient is between life and death, doctors must have other priorities than the gain of scientific information. Thus, the myth will probably remain with us for a while.

After this excursion into the sombre realm of death, let us return to our more delightful topic.

3.2 Hugging the Pain Away

We have met the French physician Michel Odent as a pioneer of natural birth in the adrenaline chapter. Furthermore he also offers his insights on painkillers, suggesting that in the case of headaches one shouldn't always reach for the drugs. "Good sex might do the trick," he assures us.

How is that supposed to work? Odent elaborates: "Every episode of our sex life is accompanied by the release of morphine-like substances. These endorphins are both hormones of pleasure and natural painkillers. During intercourse, both partners release large quantities of endorphins. Some people who suffer from migraine know that sex is a natural remedy against headaches. The release of endorphins during the copulation of mammals is well-documented."

Odent cites the example of hamsters, whose level of beta-endorphin in the blood is 86 times higher after their fifth ejaculation. While the corresponding experiment with *Homo sapiens* males has not yet been reported, endorphin release during labour and birth has also been studied with human subjects. New data have radically changed the foundations of the discussion from 40 years ago: Is the pain during birth physiologically caused, or is it the result of cultural conditioning?

Today, the concept of the physiological pain is commonly accepted, but the release of natural opiates offers a compensating protection system. This is the start of a long chain of chemical reactions: Beta-endorphins, for instance, trigger the release of prolactin, a hormone that makes a final stage contribution to the development of the baby's lungs. It is also indispensable for milk production, whereas oxytocin makes the milk flow.

3.3 Immediately After Birth, Mother and Baby are full of Opiates

"The release of endorphins during birth gives me the opportunity to emphasize that nowadays studies of pain cannot be separated from studies of desire," says Odent. He claims that there is a uniform system that can both protect us from pain and provide us with pleasure. During birth, babies release their own endorphins, such that in the hours after birth both mother and child are full of opiates. This shared "high" contributes to the beginning of the mother-child bond.

Similarly, sexual partners are full of opiates when they are close to each other. Thus, they also create a bond that follows the same model as the one between mother and child. Breastfeeding also involves endorphins, which is no surprise, as it is a crucial process for the survival of mammals, so nature has built a complex reward system around it.

3.4 Breastfeeding Makes Babies High

When a mother breastfeeds her baby, her endorphin level peaks after 20 minutes. The baby also gets a share of the rewards via the endorphins carried in the breast milk. "This is why after feeding some babies behave as if they were high," says Odent.

According to Odent, most of our knowledge on endorphins is relatively recent. It was only about thirty years ago that scientists found nerve cells sensitive to opiates in mammals. As we have explained above, this discovery led to the prediction of endogenous opiates. Nevertheless, there are still a number of white spots on the map. "Only when we have completely understood the published scientific data on endorphins will we be able to create a new basis for the better understanding of issues such as the combination of desire and pain, masochistic and sadistic

behaviour, the philosophy of suffering, religious ecstasy, and others,"
claims Odent. In his view, the love hormone oxytocin and the endorphins are part of a complex hormonal equilibrium.

In the case of a sudden oxytocin release, the resulting loving feelings
can be guided in different directions, depending on the hormonal
balance. A breastfeeding woman with a high level of prolactin, would
naturally turn her love to her baby. However, if a woman does not
breastfeed and has a low level of prolactin, she will direct the loving
feeling towards her partner.

Prolactin, the hormone required for milk secretion, suppresses sexual
excitability. If a man has a tumour that releases prolactin, impotence can
be the first symptom. On the other hand, prolactin antagonists can
stimulate erotic dreams. It is well known that in many mammalian
species the suckling mother is not available for the attention of the male.

In human cultural history, too, breastfeeding and intercourse were
often seen as incompatible. Therefore, ancient cultures looked for ways
of replacing breast milk, in order to cut the period of abstinence short.
In this context, today's milk powder has replaced the wet nurses of
earlier times.

4   NALOXONE – OPPONENT OF THE OPIATES

Naloxone (Figure 23). is a drug that blocks the effects of morphine and
opiates. Its function as an antagonist is due to the fact that it fits the
opiate receptors of the brain even better than morphine itself.

Naloxone forms a chemically favoured, extremely strong bond with
the receptor, and is even able to replace previously bound morphine
molecules. Once it has docked onto the receptor, it blocks any morphine
molecules from binding to it. On the other hand, because of subtle
differences in its shape, naloxone does not activate the receptor.

One could describe its effect to that of putty in a lock. It stops
key molecules from entering, but is unable to operate the lock itself.

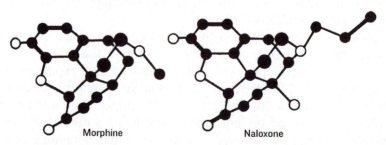

Morphine   Naloxone

Figure 23   *Morphine and naloxone*

Therefore, naloxone serves as a an antidote in opiate poisoning. Conversely, it can provoke withdrawal systems in those who are addicted to opiates and can thus be used as a diagnostic to prove addiction.

Naloxone acts as an antagonist not only against morphine, but also against the endorphins produced in the brain. "When normal people are given naloxone in a stress-free situation, they do not experience any effect," reports Robert Ornstein. Nevertheless, experiments in which volunteers are required to get a naloxone injection every day appear to be difficult to manage. "They just don't come back after the first session," the biologist complains. Even though they do not report any unpleasant effects, their behaviour allows the researcher to conclude that they don't like the drug.

4.1 Without Endorphins, The World of Intensive Emotions Remains Locked

With the help of naloxone, the US pharmacologist Avram Goldstein investigated the effect of endorphins in the brain on the perception of pleasure. He asked a group of medical students, half of them injected with naloxone, the other half with a placebo, to listen to their favourite music.

The result was baffling. While the members of the placebo group reported pleasurable effects such as shivers running down their spine, the naloxone group suffered a significant loss of their musical enjoyment.

When two young people meet for the first time and fall in love, they may also feel shivers running down their spine. Bianca and Michael know this feeling too, and they are over the moon when, after some time of separation, they can re-live this intensive happiness, thanks to their endorphins.

Having covered the chemistry of endogenous opiates, we will now briefly discuss narcotic drugs. As a comprehensive treatment of this topic would require a volume of its own, we refer to the reading list in the appendix for any further information. Here we shall restrict ourselves to those plant-derived or synthetic drugs that – thanks to their molecular architecture – are able to fool our opiate receptors.

5 NATURAL OPIATES

Let us therefore return to opium, which has given the whole class of receptors its name, and for good reasons. For millenia it has been produced from the capsules of the opium poppy (*Papaver somniferum*). Its effects are due to a number of chemical compounds known as opium alkaloids. These include morphine, which has the strongest psychogenic effect, and the relatively harmless codeine, which is found in cough syrup to this day.

All the opium alkaloids and all other substances that trigger the opium receptors in the brain are collectively known as opiates. Apart from the natural opiates, the group also includes the semisynthetic substance heroin, and the synthetic fentanyls, along with a few more substances of different chemical structures, which have only one thing in common, namely the interaction with the same type of receptors in the brain.

Opium itself, although a strong sedative, is not that harmful if taken appropriately, and only has a moderate potential to induce addiction. It is a completely different story if its main ingredient, morphine is taken neat. Since it was first isolated at the beginning of the 19th century, more and more cases of addiction were reported.

When scientists became aware of this problem, they began to search for a new drug that would hopefully be a similarly efficient painkiller without the danger of addiction. In this endeavour, they came across diacetylmorphine, which became widely known under the name of heroin. While heroin is produced in the laboratory, it does not count among the synthetic opiates. As it is based on the natural compound morphine, it is classified as semisynthetic.

Initially, great hopes were placed on heroin, which was seen as something of a miracle cure for many ailments. Only around 1930, much too late, it became clear that this drug was even more dangerous than morphine. Today, the possession of either is illegal in principle. However, as morphine has remained unique in its effectiveness as a painkiller, practitioners can still apply it in serious cases under strict limitations.

6   SYNTHETIC OPIOIDS

Apart from the classical opiates like morphine and codeine, there are a few purely synthetic substances which show similar effects as morphine, because they bind to the same types of receptors in the brain. Strictly speaking, they are not included in the definition of "opiates" (which covers only the natural opiates and their derivatives), but they fall under the more general term "opioids," which covers everything that binds to an opiate receptor. The most widely known representative of this group is methadone, which doctors in many countries are allowed to prescribe in order to assist heroin withdrawal.

However, its application remains controversial, as it also bears a high potential for addiction. Part of the idea is that methadone can be taken orally, such that junkies can be weaned off the injections with their specific risks. The desired effect is to suppress the withdrawal symptoms and to allow the drug users to return to a normal life.

During the 1960s, the fentanyls were introduced in Europe as a new class of analgetics (painkillers). For a while, these drugs were very popular, as it turned out that some of them were a thousand times more effective than morphine, implying that only very small doses were required. However, they also have their dangers, and the medical use soon declined drastically.

Their extreme potency is due to their particularly strong affinity to the opiate receptors in the brain. However, this makes it very difficult to find the right dosis of fentanyls, and this problem can easily lead to fatal overdoses. The intoxication with these compounds is often perceived as a sleep-like state, as many sensory perceptions are simply switched off by the drug. On the black market, fentanyls are sometimes sold as heroin replacement or as fake heroin, and sometimes added to low quality heroin. Regardless of such criminal activities, they are still the analgetics of choice for certain indications to this day.

6.1 Synthetic Opiates were Meant to Knock Out Terrorists

As you may remember, in October 2002 a Chechen terror commando took several hundreds of hostages during a performance of a musical in a Moscow theatre and demanded the immediate withdrawal of Russian troops from Chechnia. On the third day, Russian special units took the theatre by assault, but more than 100 hostages died in the process.

In preparation of the assault, a chemical that remained unidentified at first was distributed throughout the building via the ventilation system. Breathing in the substance was meant to lead to immediate unconsciousness. Witnesses reported that a smell was perceived very briefly, but that the effect set in so rapidly that the hostage takers had no time to either put on protective gear or use their weapons. In order to achieve the desired effect with the terrorists, the task force had obviously chosen a very high dose of the substance. Thus, acute poisoning led to the deaths of many hostages who died of respiratory or cardiac failure either at the theatre or later on, in medical care.

For a long time, Russian authorities kept quiet about the nature of the chemical that was used in the fateful assault. By now, however, it has become known that it was a fentanyl derivative. Traces of halothane found in some survivors suggest that the Russian anti-terror forces had dispersed a solution of fentanyl in halothane via the ventilation system.

If this analysis is correct, however, it throws up the question why the Russian authorities didn't inform the emergency medical staff in time. If the helpers had been equipped with naloxone as an antidote for the immediate treatment of the victims, there would have probably been fewer casualties.

7   ARE OUR OPIATE RECEPTORS A BLESSING OR A CURSE OF NATURE?

Our little excursion into the chemistry of synthetic opiates has shown very clearly that the blessings of synthetic opiates are very closely linked to their curses. As we have seen in this chapter, endorphins are a gift from Nature that we should be grateful for, as they allow us to maintain our mental balance, both in times of harmony and togetherness, and in periods of coping with pain or stress.

On the other hand, it cannot be denied that Nature has at the same time created a gate for synthetic substances with an enormous potential for addiction. But even these are useful, as they are also the most powerful painkillers known to modern medicine.

An opiate docks to its receptor, a key fits into the lock. It is up to us to make sure that the misuse made possible by false keys is kept at bay.

CHAPTER 14

Chemistry for the Eye of the Beholder
– Lipsticks through the Ages

"We have to tell every woman very clearly that painting doesn't benefit
the mouth or the lips. Moreover, there is no man who does not abhor the
idea of kissing a pair of painted lips."
Lola Montez, actress, mistress of King Ludwig I of Bavaria
"Lipstick is power."
Barbara Follet, Labour MP

1 BIANCA'S RED LIPS GET MICHAEL ALL EXCITED

Bianca tends to use lipstick very sparingly. She only applies it rarely to
go out in public, but for a private tête-à-tête, or for other special
occasions, she likes to give her sensuous lips a vigorous rouge. Michael
approves of that, and he is always delighted when Bianca applies make-
up just for him. Like many others, he associates the colour red with
passion, including the erotic kind.

As you can see from the quotations above, there have been conflicting
opinions about the lipstick, especially in times when it was an expensive
rarity and not yet a mass product. But even today, the erotic effect of
painted lips can still cause offence. Thus, in 1996, the Malaysian
parliament passed a regulation making the use of bright lipstick in
public a punishable act. It induces illegal intercourse, the reasons stated.

1.1 In the Ice Age, Ochre was all the Rage

It is impossible to tell when humans began using colours to change their
outward appearance. It is certain, however, that the use of inorganic
pigments can look back on a very long tradition. The earliest known
examples, the cave paintings of Lascaux and Altamira are around 25,000
years old. The pigments used – manganese oxides, soot, and ochre – are
thought to have had symbolic or even magical significance. Red is

117

associated with life-sustaining power. For this reason, the dead were buried with red ochre. This pigment also served ritual body painting and occasionally for make-up products, presumably to be used by the fashion-conscious members of the ice age in-crowd.

The earliest archaeological finds of real make-up utensils are from Sumer and date to the time of 3,000 BC. However, there is no indication as to whether and how the Sumerian ladies painted their lips. A bit more enlightening is the Egyptian papyrus from the 14th century BC that shows a woman retracing her lips. It suggests that the pigment was first applied to the end of a reed, which was then used to paint the lips. Archaeologists discovered further evidence from the same era in the tomb of the famous pharaoh Tutankhamun. Apart from small jars with creams and rouge, they found make-up for the lips. As we know today, it consisted of a mixture of red cochineal dye and the pigment ochre.

However, painted lips only became really fashionable in the 18th century. Starting from Paris, where it is said that even nuns indulged in it, the habit started to spread across Europe. Even severe restrictions could not stop the wave. A bill passed by the House of Commons in 1770, attempting to ban women from painting their lips red, fell flat. It even threatened women who seduced men with the help of make-up with a prosecution for witchcraft, but to no avail.

1.2 A Magic Wand of Cosmetics – Thanks to Chemistry

The lipstick as we know it today has only been in existence since 1860. It was originally invented by the German Karl Meyer, who anticipated its use for actresses on the stage only. At the beginning of the 20th century it suddenly became popular. The breakthrough came with the metal tubes, which could be made in large quantities and equipped with a mechanism to move the fragile stick up and down. Before that, the sticks came wrapped in thin tissue paper.

Over time, all the ingredients came together which have turned the lipstick into a "magic wand" of cosmetics. Roughly speaking it is composed of oils, fats and waxes. For instance, ricinus oil is found in many lipsticks, as it achieves firm binding of the pigments and gives the film on the lips strength. Bees' wax, which is also widely used, guarantees that the colour actually sticks to the lips.

Even though natural products are still the most important basis for the lipstick, chemical ingredients have also contributed added value. Take silicones, for example. Silicone waxes make the lipstick firm, silicone resins make it waterproof, and silicone oils make it feel pleasant on the skin. Together the silicones ensure that the pigments stick to the lips.

The recently fashionable "wet look" shows how silicones can set a trend. Silicones with fine-tuned refractive index make sure that the lips are not only colourful but also shimmering as though they were humid, which gives a "sensuous" impression.

The results of market research are clear: Around 92 percent of female consumers in the industrialised countries use lipsticks, which also appear at the top of the league table of items most frequently shoplifted. Other statistics reveal that Ms. Average puts around three kilograms of it on her lips over her lifetime, which is no wonder, as she appears to paint her lips more frequently than she brushes her teeth. US households are more likely to contain lipstick than mustard, and the nationwide spending on lip makeup over a year would pay the president's salary for a century.

Obviously, there are many other intriguing things to report from the interface of make-up and chemistry. However, as this would digress too far from the main topic, the chemistry of love, we will end our little excursion into the"chemistry for the eye" here and turn our attention to other ways of enhancing the sensual experience.

CHAPTER 15

Secret Scents of Seduction

"Women use perfume, because it is easier to seduce the nose than the eyes."
Jeanne Moreau (actress)

1 COME ON LET'S GO

Bianca and Michael are still at the airport, sitting in the "Zeppelin" cafe, although they have long finished their cappuccinos. "What are we waiting for?" Michael is keen to leave.

"This," replies Bianca, taking a small blue flask from her handbag. "Look, I got that from the duty free shop in San Francisco."

"A new perfume?" Michael takes the flask and reads the label.

"Yes," says Bianca, "I had a bit of time left at San Francisco, so I tried out a few of them." She laughs. "I thought this one was quite nice."

"Wait." Bianca opens it, moistens two fingers, and draws a line of scent from her earlobe down her neck.

"Mmmmm – great." Michael sniffes the scent, his eyes closed.

"I thought so," Bianca replies. "You do like it, don't you?"

"No," says Michael, smelling once more, "it turns me on."

Bianca takes his hand. "In that case . . . come on, let's go!"

When the two of them leave the Zeppelin, they are already completely immersed in their own world, and don't even notice that Marco, the friendly waiter, is waving good-bye.

1.1 From the "Dialogue with the Gods" to Modern Perfume

What just happened between Bianca and Michael is not exactly a new phenomenon, as the ancient cultures already appreciated the stimulating effects of perfume-like substances. The origins of perfume, however, are lost in the dawn of human history and cannot be traced. There is evidence for perfume use as early as 4000–5000 years ago.

The word "perfume" goes back to the Latin expression "per fumum," which means "through the smoke". This expression refers to an ancient tradition which can be traced back through millenia of history. Long before the Romans, the Sumerians, living in the area that is now Iraq, talked to their gods "through the smoke". Perfumes, the recipes and preparation of which were strictly reserved for priests, were set alight and were thought to allow the gods to descend to Earth through the smoke.

At around the same time, the Egyptians discovered perfumes. For them, as later for the Greeks and Romans, the scent was no longer reserved for the gods. Anybody could use as much perfume as they could afford to buy.

One of the most famous perfume users in ancient Egypt was Cleopatra (Figure 24). Reportedly she came up with lots of fragrances in her bid to seduce Marc Antony. On one occasion she is said to have lined the floor

Figure 24 *Cleopatra*

with rose petals and applied a mixture of jasmine oil, rose oil and honey to her body.

On festive occasions, some wealthy Egyptians liked to place a cone of ointment on their heads. In the course of the proceedings, the fats in the cone would melt, run into the wig, and soak it with scents, thus guaranteeing a long-lasting fragrance. At Egyptian funerals, it was indispensable to embalm the body of the deceased with strongly scented resins and oils. According to ancient Egyptian beliefs, the perfectly conserved body was a prerequisite for life after death.

While these early examples of perfume use were closely linked to religious practices, this aspect was slowly replaced by its use as a luxury object. Over millenia, caravans travelling through deserts and across mountain ranges, and ships crossing the oceans transported the ingredients required to make perfumes between continents. Fragrant essences, precious and rare herbs and spices, highly valuable animal products like ambergris, musk, and civet were among those goods. A bewitching fragrance became a symbol of power, wealth and excess. The triumphant advance of perfume through the centuries became unstoppable.

Perfumes as we know them today – mixtures of alcohol and essential oils – were first produced in the 14th century, following the successful destillation of concentrated alcoholic solution from wine around the year 1200. Adding alcohol to the mixture helped to preserve the sensitive natural products involved. The new fragrances were also seen as cures against all kinds of ailments and infections. In the 15th century the Italians – then the richest people in Europe – witnessed a triumphant success of perfumes. Especially in the merchant republic of Venice they were used lavishly.

From Italy, perfume making spread primarily to France. Under Louis XIV, its use became popular there. At the end of the 18th century, the first "Eau de Cologne" came to the market, to compete with the heavier fragrances of the past. To this day, it is produced according to a recipe of the Italian Giovanni Maria Farina.

After the French Revolution, the use of perfumes was frowned upon. Only Napoleon, who is said to have had a penchant for the light Eau de Cologne, brought perfumes back from the wilderness. To this day, France is the heart of the perfume industry. Specifically, the town of Grasse, where fragrant plants have been cultivated for centuries, has made its name as the capital of perfumes.

Today, using perfume is commonplace and a part of everyday life. The fragrant whiff of luxury completes the well-groomed appearance of both men and women. Thanks to today's opportunities to create a plethora of synthetic ingredients, there are almost no limits to the imagination of the perfume-makers. There are around 32,000 different substances at their disposal, of which an experienced master of the trade can smell around 3,000 apart.

1.2 Top Notes, Middle Notes, and Base Notes: An Excursion into the Chemistry of Fragrances

In principle, a perfume is a melody composed from different notes of fragrances. The music of the scents is a composition of various plant, animal, and synthetic substances dissolved in alcohol. Depending on their effects, these are grouped into top, middle and base notes.

- The top note. The most volatile ingredients of a perfume determine its top note, the first fragrant chord of the composition. The top note is the whiff of fragrance that you notice when you open a flacon or immediately after applying the perfume. Because of its volatility, however, this fragrance is very short-lived. Perfumers often use citrus oils like bergamot, clementine, bitter orange, or lemon.
- The middle note. It develops later than the top note, but it marks the main emphasis of the fragrant composition, so it is the "theme" of the perfume. The middle note typically unfolds within an hour and can last up to ten hours. It is often based on heavy, sweet flower fragrances, such as jasmine, lily of the valley, lilac, carnation, or rose.
- The base note. It is characterised by the least volatile ingredients and gives the fragrance depth. The smell can remind of wet soil, oriental spices, moss, or leather, or even be animal-like. Accordingly, this is the note which allows us to recognise people by their smell, and which conjures up memories. Typical base notes include sandalwood, vanilla, amber, oakmoss, and vetiver oil, which is extracted from a tropical wood and smells of soil and wood.

Most of these natural fragrants, however, are extremely expensive, and have therefore been replaced by synthetic ingredients. This may sound disillusioning, but if it weren't for chemical synthesis, perfumes would remain beyond the financial reach of most of us. For example, the production of one kilogram of attar of roses requires five tons of petals. This is why it costs well over 3,500 pounds. As a comedian once remarked, "chemistry is like nature, only cheaper."

Indeed, chemistry has succeeded in producing a number of compounds that closely resemble the natural fragrances. Take for example the esters, readily formed by reactions of aldehydes with carboxylic acids. You may be familiar with the fruity-fresh smell of acetic ester. However, aldehydes are more important to the perfume industry.

Aldehydes are compounds obtained from alcohols by removal of two hydrogen atoms (oxidation) and can trigger very diverse odour perceptions. While some of them correspond to natural products, others represent new fragrances that cannot be found in Nature.

Readers with laboratory experience might sniff at the idea of using aldehydes in perfumes, as they will remember the unpleasant, biting smell of the most volatile representatives of the group, formaldehyde and acetaldehyde. However, things change when you add more carbon atoms. Aldehydes with 10 to 14 carbon atoms tend to smell nicer, and this group includes both undecanal, which we will meet as a pheromone in the next chapter, and 2-methyl undecanal, the top note of the perfume classic Chanel No 5.

How to describe the smell of 2-methyl undecanal? "The smell resembles that of incense and pine wood," claims the trusted Römpp encyclopedia of chemistry. When Marilyn Monroe was asked what she wore in bed, she famously replied "Only Chanel No. 5." In the prudish climate predominant in the USA in the 1950s, this innocent quip was sufficient to whip up a scandal. A scandal ultimately caused by a simple aldehyde, whose name the actress probably was blissfully unaware of.

1.3   A Frenchman had a Nose for it

Let us now proceed to the high priests of fragrances. The Frenchman Francois Coty is widely regarded as the "Father of modern perfumes." He developed many synthetic fragrances and implemented important technical improvements. Because he literally "had a nose for it," Coty soon became one of the wealthiest men in France. He was a senator in Corsica, an art collector, and finally the publisher of the newspaper Le Figaro.

His entry into this career was just as unusual. Back in 1904, he approached the head of the purchasing department at a prestigious Parisian department store, in order to persuade him to include his home-made fragrance in the store's range. As his request was met with a decisive "non," Coty is said to have dropped the vial on the floor. Any resentment over the spillage must have evaporated quickly, as the fragrance wafting through the aisles soon attracted eager customers who crowded around the two men.

Coty's name is often heard in connection with a fragrance known as "chypre." In 1917, Coty developed this composition based on a lichen found in Cyprus. With it he created a new category of fragrances. Today, chypre is a collective name for a group of perfumes, which are characterised by the combination of a fresh top note with a base of oakmoss, labdanum, and patchouli. Many warm, erotic, and sensuous perfumes are found in this group.

Throughout the perfume industry, Coty is seen as a genius. From 1904 through to his death in 1934, he developed no less than 50 successful perfumes. Immediately after his death, another blockbuster came to the market: "vanilla fields." The perfume targeted at a broad range of customers soon became the best-selling fragrance in the USA.

1.4 Animal Instincts

Apart from the manifold plant products used in perfumes, there are also a few fragrances derived from animal secretions. Only four different fragrances of animal origin have become important for the perfume industry:

- ambergris (a pathogenic secretion of sperm whales),
- castor (the sexual secretion of beavers)
- musk (the sexual secretion of the musk ox)
- cibet (the marking secretion of cibet cats of both genders)

As you can imagine, the undiluted smell of these substances is everything but enticing.

Ambergris secreted by sperm whales is sometimes found floating on the oceans. Reportedly, some people became rich by visiting the right beach at the right time and finding large amounts of the substance. A byproduct of whaling, ambergris can be obtained from dead whales, but then it has to be kept in sea water to mature for around a year before it can be used for perfumes.

To harvest musk and castor, the animals had to be killed. Only cibet can be obtained from live cibet cats. By now, however, perfume makers can leave all these animals in peace, as their secretions have been replaced by similar synthetic substances.

1.5 Take a Deep Breath – How Our Sense of Smell Works

"Whoever looks closely at the nose of their beloved through a magnifying glass will discover hairy mountains that will make them shudder." We assume that Johann Wolfgang von Goethe was joking when he wrote these lines, and that he didn't have any plans for further research into the anatomy of our olfactory organ. We, however, will do it without shuddering, as research in this area has brought to light some interesting results related to the evolution of our sense of smell.

Although the interest in olfactory perception goes back to Aristotle, who saw it as an intermediate between human and animal senses, it has remained a poor relation of sensory research over centuries. Only in the last few decades it has increasingly arrived in the limelight.

As we know today, a person has around 30 million olfactory cells, which are located in the olfactory epithelia, on the ceiling of the nasal cavities. These epithelia consist of a layer of basal and supporting cells, from which the up to 30 million olfactory cells protrude into knobs. Each knob is equipped with around five very fine sensory hairs, known as cilia. The olfactory cells have the task of translating any odours they detect into electrical signals. However, up to now it has not been elucidated in which ways exactly this signal conversion occurs.

What is clear is that the olfactory cells are capable of sensing odour molecules. When that happens, they create a nerve impulse, which the olfactory nerves transmit to an outstation of the brain, no larger than a matchstick head, known as the olfactory bulb.

Thus, unlike visual and auditory signals which travel via the thalamus, smell goes directly to the oldest and least investigated part of the brain. This is why the overwhelming majority of our odour perception occurs on a subconscious level, so we are unable to control it consciously.

This finding is consistent with everyday experience. Maybe you have had the following experience as well. You are on a train, in a restaurant, staying with friends, or in a hotel room, and you suddenly notice a familiar smell, but you can't place it right now. Maybe you'll never figure out what it was and where you smelled it before. If you do find the connection, it is often linked to specific places, people, or events.

A French medical dictionary of the early 19th century described the sense of smell as a "sense of tender remembrance." Conversely, it is almost impossible to recall a smell into our memory. Even smells from early childhood are anchored in our memory, but we only recall them if we encounter them again.

1.6   Is the Gradual Loss of Our Sense of Smell an Error of Evolution?

Smelling is one of the oldest sensory perceptions, but in primates and specifically in humans the talent for it has obviously declined rapidly in the course of evolution. This is at least the conclusion reached independently by the groups of Doron Lancet at the Weizman Institute at Rehovot, Israel, and of Svante Pääbo at the Max Planck Institute for Evolutionary Anthropology at Leipzig, Germany.

Pääbo is regarded one of the founders of palaeogenetics. "This is an approach aimed at elucidating the history and origin of mankind using genetic variation," he explains. One of the experimental approaches he uses is the analysis of ancient samples, *e.g.* from cave bears, mammoths, or Neanderthals. Alternatively, one could compare the genome of today's humans with other primates.

Pääbo, who was born in Sweden to Finnish parents, created a stir recently, when he compared DNA from the bones of a Neanderthal to the genetic material of humans and chimps. The differences observed both ways showed that Neanderthal was a separate lineage and not an ancestor of homo sapiens.

As we have seen in Chapter 4, chimps and humans differ only by around one percent of the letters in their genetic makeup. Even with the 99% agreement, this leaves around 30 million differences. However, not every difference will have an effect, as genes may be switched off. Only the functional genes account for the differences between species. It's the differences in functional genes that geneticists like Svante Pääbo are trying to find.

Obviously, the differences between apes and people are most conspicuous in the cognitive area. Pääbo's group could demonstrate that the human brain has changed more than that of the chimp since the time when the species diverged. These differences have enabled humans to make the best use of the genes related to intellectual functions.

The evolution of olfactory perception tells a different story. With around 1000 genes it may be the most lavishly equipped of our senses, but this equipment is going to the dogs, as two thirds of the genes are no longer functional. Pääbo has more bad news: "This process is not finished yet, as the ability to identify and distinguish smells declines further in humans."

The causes of the decline are not clear yet. Together with the Weizman group, the Leipzig researchers have proposed the hypothesis that it may be coupled to the improvement in colour vision. Our ancestors, as the theory goes, were no longer forced to identify ripe fruit or berries by their smell, as they were increasingly able to distinguish their blue or red colours from the green background.

Thus the researchers believe that the improved ability to process visual stimuli in the brain made smell dispensable. For instance, we are better at distinguishing colours than other mammals that are better at sniffing. Moreover, we have learned to distinguish fellow human beings by looking at their faces. Recognising their smell became less and less important.

1.7 Show Your True Colours and I Tell You Which Fragrance You Prefer

No matter how individualistic a woman may be and how many different perfumes she uses, most probably she is going to stick with one kind of fragrance. For example she might prefer fruity perfumes, or she may have a penchant for chypre notes. This general preference mirrors her image of herself, *e.g.* active and dynamic, or romantic and mysterious. The fragrance is a mirror of her personality.

That's all good and well, you may think, but what does it have to do with colour vision? Quite a lot, as it turns out. Because psychologists have known for a long time that our favourite colours can reveal a lot about us. They express our feelings and moods. Perfumers have found out that our favourite colours are also linked to the fragrances we prefer. They carried out the so-called colour rosette test, which revealed that the personality types known from colour psychology – extrovert, introvert, emotionally changeable, emotionally stable, with various intermediate forms – can be matched with just as many clearly distinguishable fragrance requirements. In other words: our favourite colours reveal which fragrances we like.

This is important for lovers, as love appears to work only when our noses agree. As we will find out in the next chapter, there is more to olfactory compatibility than Marilyn Monroe's Chanel No. 5, as unconscious perception of pheromones appears to play an important role. Once the noses are satisfied, however, visual input becomes important (see Box below).

COLOURS CAN BE ENTICING

Our appetite for a delicious meal often depends crucially on how it is presented. We enjoy it with our eyes as well, not just with the taste buds. If we tell you that the same applies to our sexual desires, you may scold us for stating the blindingly obvious.

However, the group of Timothy D. Smith at the University of Michigan at Ann Arbor wanted to dig deeper and find out why this is so obviously true. Why, the researchers wondered, does the *Homo sapiens* male have a tendency to be stimulated by visual signals, *e.g.* reading a shining colourful bikini as an invitation to start the mating game?

This example also illustrates a common misunderstanding between the sexes. Most probably, if the bikini wearer aimed at covering rather than emphasizing her assets, her refusal will be harsh and the over-eager male may end up bitterly disappointed.

At least the male desire triggered by optical stimuli can be explained by evolution, the researchers from Ann Arbor claim. In the course of human evolution, our ancestors gradually learned to trust their eyes more than their noses.

The researchers think that this situation arose because we and our primate cousins only have very limited abilities to respond to chemical signals in mating. Although the enticing fragrance of the opposite sex has not yet lost all of its effect, humans and great apes have lost crucial genetic elements linked to its perception. The new world monkeys, whose lineage separated from ours some 35 million years ago, can still rely on their noses when choosing a partner. In humans, however, many of the genes have degenerated to pseudogenes.

While pseudogenes are longer dismissed as junk DNA and may have certain functions, it is unclear whether they play a part in maintaining what little remains of our sense of smell, which still allows us to detect some kinds of chemical signals, as we will discuss in the next chapter.

It is established, however, that humans possess only around 400 functional olfactory genes, while chimps still have 700. In contrast, typical new world monkeys such as the squirrel monkey or the black spider monkey have an excellent nose, but are colourblind. Consequently, they sniff at each other before mating, rather than looking at each other.

A "missing link" between these extremes is the howler, which has both abilities. So far, Smith and his co-workers have no satisfactory explanation for this double act. Maybe it's just the exception to the newly discovered rule.

If the general rule is that you get either colour vision or a good nose, we should not complain about our fate. Indeed, men who spend their free time by the swimming pool making eyes at women in conspicuous bikinis should be grateful. The nose of a squirrel monkey would be no use to them by the poolside, as any chemical messengers wafting their way would have been destroyed by the chlorine.

CHAPTER 16

Pheromones – Words in the Dialogue of Fragrances

"Fragrances are like music for our senses."
Ancient Persian proverb

Moths are famously attracted to the light. However, there are far more efficient attractants, which will exert their effect over distances of kilometres. These are Cupid's chemical arrows, also known as pheromones. A few molecules of these chemical attractants will be sufficient to drive a male animal crazy. This chemical body language is widespread in the animal kingdom. For example, pigs, mice and frogs use it. Even yeast cells send chemical love letters.

Biologically speaking, pheromones are defined as chemicals produced and released by living organisms which in some way influence the behaviour of other individuals of the same species. They have been studied mainly in insects, where they are the basis of a complex communication system. Depending on the task they fulfil, one distinguishes between the following groups:

- Aggregation pheromones serve the cohesion of a group.
- Marker pheromones mark food sources and homes.
- Alarm pheromones warn of dangers.
- Sex pheromones bring the sexes together and thus help to propagate the species.

Pheromones are typically secreted from glands in the female animal and perceived by the male with the help of specific receptor organs. As they typically occur only in extremely small amounts, they have initially been difficult to detect and to study, but thanks to modern analytical techniques, we now know the chemical structures of a large number of them. Often they are relatively simple molecules, which are easy to synthesize and have therefore become alternative means of crop protection.

130

As pheromones are known to have a decisive influence on insect behaviour, they can be used to keep insects away from any crops they might otherwise damage. For example, placing sex pheromones in commercially available insect traps can increase the success rate significantly. Other methods aim at disrupting or at least disturbing the communication system of the animals.

Anybody who thinks that this "sex trap" is a mean trick should consider the precedents in Nature. For example, the orchid *Ophrys sphegodes* produces long-chain hydrocarbons that attract the males of a specific kind of solitary bee (from the species *Andrena nigroaenea* or *A. limata*), ensuring the plant gets the desired pollination service.

Analysing the chemical messengers, researchers were surprised to find that *Ophrys sphegodes* can mimic no less than 14 out of the 15 sex pheromones of those bee species, and that it even matches the ratio of their concentrations to the original. This mimicry seduces the male bees to follow the plant fragrance blindly and to get involved with the flowers instead of the females of their own species. For this strange and fascinating interplay between flora and fauna evolutionary biologists created the term "pseudocopulation."

Other researchers set out to investigate whether the highly sensitive olfactory organs of insects might be useful for the development of a new generation of biosensors. A prime example of this high sensitivity is the moth *Manduca sexta* (Carolina sphinx). The male animal carries around 40,000 sensory cells on its antennae, making sure that no female pheromones waft by undetected. A research group at the University of Kaiserslautern, Germany, has succeeded in isolating the sensory structures from the antennae and record their electrical output in the laboratory.

Surprisingly, the insects can still detect pheromones at concentrations 1,000-fold lower than the detection threshold of conventional chemical sensors. This is why classical chemical methods mostly fail to pick up pheromones, which are typically as dilute as a milligram of substance dissolved in a swimming pool.

1   DO PHEROMONES MAKE HUMANS HORNY, TOO?

Whether or not pheromones play a role for our species has been controversial for a long time. Even though the debate hasn't quite subsided yet, more and more scientists believe that we have a chemical sensor on top of our five classical senses.

"At the origin of all life there was molecular recognition," emphasizes Wittko Francke, a professor of organic chemistry at the University of Hamburg, Germany. Therefore, he argues, chemical communication existed long before there were any higher organisms, let alone mammals. Even though the chemical communication channel does not play such a crucial role in humans as it does among animals, it does spring into action after the phase of getting to know each other.

In other words, even though we mainly communicate verbally and visually, we are able to send out molecular fragrance messages and to receive them as well. Pheromones could be called the chemical words in the dialogue of the fragrances. To this day, they have remained mysterious to us.

HOW EGG CELLS ATTRACT SPERMS

How do sperms find the way to the egg cell? Scientists in Germany have found new answers to this old question. With specialised techniques, they could observe what happens in the first few milliseconds after a sperm has come into contact with the attractant of an egg cell. The biophysicists found out that for each sperm, a single molecule is sufficient to trigger a signalling chain. A comparably high sensitivity has so far only been observed for visual cells, which can be excited by a single quantum of light (a photon).

Egg cells release chemical messengers to attract sperms. The sperms use the resulting gradient, *i.e.* the direction in which the concentration of this attractant increases, for their orientation, and ultimately to locate the egg. The control of swimming behaviour by a chemical excitation – a biological phenomenon known as chemotaxis – is observed in the sperms of a wide range of organisms from marine invertebrates through to *Homo sapiens.* Until recently, however, it took several seconds to observe how the sperms respond to the attractant.

But what happens in the first milliseconds after the sperm has encountered the attractant? The group of Benjamin Kaupp and Ingo Weyand at Jülich, Germany, has studied how the chemical excitation is processed, using rapid mixing of see-urchin sperms with a short chain polypeptide attractant. Researchers at Berlin had modified this peptide in such a way that it remained inactive initially (also known as a "caged" molecule). By means of an UV flash, which triggered a photochemical reaction in the peptide, it could be activated in an instant, during the experiment.

With this trick, the researchers could control precisely the moment from which the sperms were exposed to the attractant. They could observe how the peptide initially binds to a receptor protein on the cell surface of the sperm. This event then triggers the synthesis of an internal messenger molecule, a cyclic nucleotide known as cGMP (for cyclic guanosine monophosphate).

The researchers found out that the concentration of this messenger molecule within the sperms increases very rapidly after the encounter with the attractant. The cGMP molecules in turn trigger the opening of certain ion channels (like those we discussed in the chapter about the brain), which specifically allow calcium ions to enter the interior space of the sperm cell.

In the absence of the attractant, the sperm's tail beats regularly and propels the cell forward on a spiralling path. Using a microscope, the researchers observed how the sperms change their swimming behaviour when they get in contact with the attractant. After the calcium ions have entered the cell, the tail begins to move more asymmetrically. The sperms carry out a U-turn. Ultimately, they gather round the source of the attractant, *i.e.* the egg cell.

A single molecule of the attractant is sufficient to open the ion channel in the sperms and let calcium ions enter the cell. At the moment, the researchers are busy investigating further details in this signalling chain, plugging the last gaps in our understanding of the series of events in reproduction.

But what is the molecular nature of this potent attractant? Researchers led by Marc Spehr and Hans Hatt at the University of Bochum, Germany, could show that it is identical with the odorant bourgeonal, whose fragrance is reminiscent of lily of the valley – a note that obviously attracts both lovers and sperms.

In addition, the researchers discovered another fragrant molecule that binds to the same receptors, but blocks their activity in competition to the bourgeonal. This blocker is the aldehyde known as undecanal, which is found in citrus oils and used in the perfume industry. In its presence, the lily of the valley fragrance loses its effect. But how does the inhibitor influence the swimming behaviour of sperms?

Comparative experiments showed that human sperms normally orient themselves directly towards the lily of the valley fragrance and furthermore double their speed. In the presence of the blocker undecanal, however, they were swimming slowly and without a directional preference. Further research will be directed towards the use of this antagonist in the development of new means of contraception.

1.1 A Relic from Earlier Days of Evolution?

David Berliner, the US researcher who in 1963 surprised the scientific community with the discovery of human pheromones, is one of the scientists who believe that the solution to the pheromone mystery will be found in the so-called vomero-nasal organ (VON: from vomer, a small bone located in the middle of the nose), also known as Jacobson's organ, after Ludwig Levin Jacobson who discovered it in humans at the beginning of the 19th century. Over many years, the VNO was considered an inactive relic from our evolutionary history, as its function had been established only among animals, where it is credited with the "sixth sense."

In the human nose, it is represented as a minuscule epithelial tube that found very little attention. Surgeons operating on the nose did not even remotely consider preserving this "useless" little thing.

New research results, however, show that this attitude needs to be revised. "Today we know that the VNO also plays a role in humans, but its function has not been completely elucidated yet," says Thomas Hummel, who chairs a research department for "taste and smell" at the University of Dresden, Germany.

"Across all ages, we only discovered the VNO in 60–70 percent of the over 180 people we looked at," Hummel explains. Often, the researchers only found it on one side of the nose, and not on both as one would expect. Tissue characterisation revealed typical aspects of a sensory organ. "However, we could not find any nerve connections, as one would expect in such a case," Hummel continues. All in all, his results appear to be consistent with the theory that the VNO is an inactive relic from our evolutionary history.

Remarkably, an embryo's VNO grows very fast in the 20th week of pregnancy, so it is quite prominent in late development and in newborns. Some researchers suspect that this "sixth sense" even enables infants to sniff out their mothers.

Gerd Kobal, a physiologist at the University of Erlangen-Nürnberg, Germany, works on the assumption that the VNO is still very active in babies, but that the connection to the brain is gradually lost when they grow up. Otolaryngologist Volker Jahnke from the Charité hospital at Berlin, however, doubts that it is lost completely. He argues that electron microscopic investigation of nose epithelia has shown that even in adults there must be nerves connecting the VNO to the brain.

In collaboration with the anatomist Hans-Joachim Merker from the Free University at Berlin, Jahnke removed the VNO from men and women who – for a variety of reasons – had to undergo surgery in that region. With the electron microscope, the researchers revealed a surprisingly complex form. It is a thin tube of variable length (2–8 mm) and width (0.2–2 mm), closed at one end, and obviously resulting from a part of the epithelium caving in. It is found in the lower, frontal part of the nasal septum. As it is open towards the nasal cavities, it can continuously take part in substance exchange with the environment.

There are a few conspicuous cells at the rear end of the tube, which are clearly different from normal epithelial cells. There are especially bright, elongated sensory cells, linked to numerous nerve fibres. In principle, they resemble the cells of the olfactory epithelium, but they are more closely linked to each other and have small, mobile protuberances which look almost like handles.

1.2 The Truffle Pig in us Decides Whether We Get on with Each Other

Beneath the top layer of cells, there are blood vessels and a mesh of nerves. This complex and unique structure of the human VNO, which differs significantly from those of other mammals, makes Jahnke and Merker conclude that it might have special tasks, for instance sending chemical messages via body fluids. The researchers assume that glands in the nose release a watery secretion, which dissolves the volatile pheromones in order to ship them to the sensory cells. Whether the plentiful nerve fibres pass on the signals from there to certain parts of the brain remains to be discovered.

And yet, the insights from the last few years have already influenced the way surgeons operate. If at all possible, they will try to preserve this special organ, which might still turn out to house our "sixth sense." Because it might well be that this organ is involved in the decision whether two people like each other, and which partner we choose unconsciously. Moreover, increasing evidence suggests that the role of pheromones in humans has been underestimated by far.

The desire for closeness triggered by the body's chemistry will doubtlessly travel from the nose to the brain. When Bianca met Michael and hugged him, his pheromones formed in the adrenal cortex – especially androstenol and its oxidation product androstenone – produced a smell that not only was familiar to her, but also gave her the feeling of affection, warmth and protection.

1.3 Men are Smellier Than Women

The chemical structures of both androstenone and androstenol resemble that of the sex hormone testosterone. Androstenone has a fragrance reminiscent of musk, while androstenol smells a bit like sandalwood. As men's sweat contains about six times more of these compounds than women's, men tend to be a bit smellier than women.

Nevertheless, women don't necessarily perceive the smell of a male armpit as unpleasant. On the contrary: Just before ovulation, they are said to perceive the otherwise unpleasant smell of androstenone as rather attractive. Hormone researcher Wittko Francke recalls: "Back in the old days, young men who wanted to go dancing used to tuck their handkerchiefs under their armpits for a while." Apparently, this helped to attract the ladies.

However, in some people, the natural body odours are rather repellant. The French king Louis XIV is said to have been such a case. Reportedly, his feet were so smelly that his courtiers kept their distance because they could not stand the whiff. Rasputin appears to have smelled like a goat, but for different reasons. The fragrance emanated not from his feet, but from his sperm. Allegedly, women worshipped him because they could practically smell his virility. However, it is unclear how much of this legend must be attributed to male wishful thinking.

Popular tradition also reports a number of unproven olfactory clichees: Allegedly eunuchs can be recognised by their bland, sexless odour, while bachelors smell of dried figs. Unlike "real men" who smell like musk. Well, if they don't, there are perfumes to make them, as we saw in the previous chapter.

SWEAT AND THE SECRET LANGUAGE OF MOLECULES

Men's sweat is sexy – this is the surprising result of a study conducted at the University of Northumbria at Newcastle in 2000. Apparently, women find men more attractive when they can smell their sweat and unconsciously perceive the pheromones contained in it.

The unusual experiment involved 16 female students who were led to believe that they were taking part in a test of new brands of soft drinks, while in reality the researchers were after their unconscious reaction to male sweat. The women were asked to judge the attractivity of men shown in photos. In a second round, they were set the same task, but this time, invisible to them, a cloth soaked in male sweat was nearby. The effect was "enormous," reports Nick Neave, one of the researchers involved. "In the presence of the pheromones, men were judged much, much more attractive," he summarises the results. The most surprising result is that those men who had previously been judged the least attractive by all participants, were able to catch up. In the presence of the pheromone trap, they were rated almost as attractive as the best-looking men.

According to a more recent study conducted by biologists at the University of Pennsylvania, male sweat also has a relaxing effect on women. Furthermore, the secretions are said to have a positive effect on the menstrual cycle.

In their research, the scientists took swabs from the armpits of male subjects who had not used any deodorant for four weeks. They applied the secretions to the upper lips of female participants, who promptly reported better mood and lower perceived stress. Subsequent blood tests revealed an increase in the concentration of the luteinising hormone (LH), which is responsible for the maturing of the follicles, to an extent that was similar to the natural increase just before ovulation. According to the researchers, these results might turn out to be helpful for the therapy of some kinds of infertility. However, this would require first of all to identify and isolate the effective components of male sweat.

In this experiment, too, the participants were left unaware of the nature of the substance used, believing they took part in a perfume test. "The test results suggest that there is an unconscious chemical communication between the sexes," says biologist Charles Wysocki, one of the authors of the study.

1.4 Napoleon to His Wife: "Don't Wash, Will Arrive in Three Days."

Obviously, Napoleon Bonaparte was one of those men who preferred "stronger" fragrances (Figure 25). Returning from a campaign, he is said to have sent a private message to his Joséphine by a reliable courier, which allegedly read: "Don't wash, will arrive in three days." Did he hate the idea to find a clothesline full of clean laundry on his return? Historians agree (for once) that he had something else in mind...

Many people believe that they are most attractive when they are squeaky clean and smell of nothing but perfume. But is that really true? Recently, ethologist Karl Grammer from Vienna showed that the fragrance that turns men on does not come from cosmetic products. In his study, Grammer and his coworkers got 66 young men to sniff out synthetic copulins.

Chemically speaking, copulins are a mixture of short-chain fatty acids. They are contained in the vaginal secretion, where they occur especially during specific phases of the menstrual cycle, namely before and during menstruation, and during ovulation. A control was carried out with pure water vapour. In addition, the researchers presented photos of five women rated with different degrees of attractivity. Before and after the smell exposure, the men were asked to provide samples of their saliva, which were used for the analysis of testosterone concentrations.

1.5 Copulins – the "Chemical Weapons" of Women

While in the group that only sniffed water vapour, the testosterone levels decreased slightly, the copulin group recorded an increase of hormone concentration. This response was completely independent of the

Figure 25   *Napoleon Bonaparte*

attractivity of the women in the photos. Remarkably, the ovulation copulins also had the effect of levelling the attractivity of the women in the eyes of the men. The less desirable ones gained in attractivity, with the least attractive one making the largest gain.

Apparently there is a kind of chemical warfare going on between the sexes, and it happens on a level that is not accessible to our cognitive abilities. "If you let a man smell copulins, his ability to judge the attractivity of a woman breaks down completely," Grammer concludes.

1.6   Artificial Pheromones in Perfumes make Men Keen to Cuddle

Can synthetically produced pheromones fully replace the natural ones? With copulins it appears to have worked, but how about the other pheromones? Researchers from San Francisco State University reported in 2002 how artificially produced pheromones can fool us. The test involved women of different age groups and ethnic origin, who were asked to hand in a flask of their favourite perfume.

The researchers added sex pheromones to the perfumes of one half of the participants, while the others got theirs back unaltered. Only the researchers knew who was in which group. During the experiment, the women kept a diary, recording how often they were kissed, whether their partners snuggled up to them in their sleep, and how often they had sex.

The results were clear: 74% of the women who had pheromones in their perfumes were kissed more often and longer, their partners became more cuddly, and they had sex up to six times more often than without the "special perfume." Singles recorded a sudden increase in the number of dates. In contrast, the women who had used the untampered perfumes, only 23% observed a change in their sex life.

Hormone researcher Wittko Francke recalls another sneaky experiment. "There have also been experiments with tampered textiles," he reports. Apparently, small doses of androstenone in the cloth generated significant increases in consumers' interest compared to untreated textiles.

American hormone researchers have also set their traps in dental surgeries. When two different chairs were up for choice, the majority of the (female) patients followed the chemical trace. Further evidence for the "chemical sense" of humans is the sexually stimulating effect attributed to truffles. "These fungi also contain androsterol and androstenone," explains Francke. With a smile, he adds: "It appears that not only pigs appreciate that."

Now, researchers have also shown that humans release substances to the environment which influence the metabolism and behaviour of others. The best-studied example is the synchronisation of the menstrual cycles between women who live together. Back in 1970, Wellesley undergraduate Martha McClintock noticed this effect among her dormmates and wrote a thesis on this topic. At that point, however, it remained unclear whether chemical substances or social factors were responsible for the phenomenon.

Nearly two decades later, McClintock returned to that topic and investigated it experimentally, by exposing women to the armpit secretions sampled from other women during specific phases of their menstrual cycle. Thus she could show that cycles of women who are regularly exposed to another woman's body odours will by and by synchronise with that of the "odour provider." Depending on the stage of the cycle at which women are exposed to these "hidden fragrances," their cycles will shorten or lengthen. Furthermore, McClintock reports that the substances in question do not produce a smell that we can perceive consciously.

McClintock now assumes that there are at least two different substances controlling these processes. While she is still scratching her head over the evolutionary significance of the synchronisation effect, she is already considering practical applications. According to her, this effect could ultimately lead to an innovative, but also more natural way of contraception.

1.7 "I Don't Like Your Smell, Because Our Genes are Too Similar"

There is an element of luck and coincidence involved when two people meet for the first time and fall in love with each other. Beyond that, however, researchers in Germany and Switzerland claim that the genetic disposition decides whether a couple has the right chemistry or not. According to their findings, every person has a kind of genetic fingerprint, a characteristic and innate profile of volatile substances, hidden in their body odour.

Specific molecular tissue markers, known as HLA (human leukocyte antigens) and bundled in the so-called major histocompatibility complex (MHC) are thought to be responsible for the "personal note" of a person's body odour. Moreover, almost every cell of the body carries these molecules on its surface.

The MHC molecules are part of the immune system's provision to ensure that the body can recognise its own cells and distinguish them from foreign ones. They are the reason why, after organ transplants, the recipient's immune system has to be suppressed. Like antibodies, they can be produced in millions of different varieties, defining each individual as the carrier of a unique combination of molecules.

The more different these markers are between two people, the more pleasant they perceive each other's body odours to be. Research with mice has confirmed that they, too, use HLA molecules in their choice of partners, mating strictly only with candidates whose tissue markers are different.

If you can't stand somebody's smell, that appears to be a warning sign from Nature, which in fact makes a lot of sense. Most of the time, social control makes sure that people don't marry their close relatives. However, a person might encounter a potential partner with a high degree of genetic similarity by coincidence or unaware of the relatedness. For the offspring, it is generally bad news if the parents are related or very similar genetically, as there is a smaller amount of genetic variety accessible to the new life. Variety is important for example for the immune system and its ability to fight off infections.

Research conducted at the University of Bern, Switzerland, has shown in detail how people recognise (potential) partners from the smell of their HLA molecules. T-shirts worn by male volunteers were the samples that the female subjects of the study were asked to sniff at.

In order to avoid any bias from other odours, the volunteers renounced alcohol, nicotine and garlic for the duration of the experiment. Twelve female "sniffers" gave their judgements regarding the T-shirts on a scale of sensory perception ranging from enticing through to revolting. They were also encouraged to note any memory associations with fathers, partners, or exes.

The study revealed that the female subjects felt repelled by the odour samples if the man in question had similar HLA molecules to their own. In contrast, the odours of men with complementary tissue markers were rated particularly attractive. Geneticists have also found that the genes of the HLA molecules responsible for the personalised body odour are directly adjacent to those that determine our sense of smell.

Just like the detailed structures of HLA molecules, those of the olfactory receptors in our nose are genetically determined. For the dating game, this implies that any individual fragrance can only be successful if it meets a nose with the matching receptors.

Thus, the attraction between two people is crucially influenced by odours. So, the next time you notice somebody's BO, don't wrinkle your nose at it. Instead you should appreciate its information content.

CHAPTER 17

The ABC of Aphrodisiacs

"Every girl has the right to take desperate measures,
in order to ensnare the man of her choice."
Agatha Christie (Writer)

1 A LITTLE HELP FROM MY FRIENDS

If all the hormones, pheromones, and perfumes discussed in the previous chapters don't make your juices flow, you could try one of the many substances traditionally regarded as aphrodisiacs, a selection of which we will present in this chapter. Should that fail, well there is always the little blue pill which – in the next chapter – will be our last resort and final stop in our travels through the chemistry of love.

The word aphrodisiac honours Aphrodite, the ancient Greek goddess of love. In fact, the ancient Greeks already new a number of "stimulating" plants, which they called the "choras aphrodisias," or the round dance of the love plants. In Greek mythology, aphrodisiacs were Aphrodite's potions. The Romans, too, were interested in this topic. They had a whole range of plants and preparations at their disposal, which they called "venera," attributing them to Venus.

There is probably no other field in medicine where myths and legends have so stubbornly survived across the ages as in the area of aphrodisiacs. Many of them contain no active ingredients whatsoever, or only compounds with a mildly stimulating effect.

However, there may still be a pronounced placebo effect which might drive up the libido. Conceivably, this effect may even be stronger in aphrodisiacs than in other putative drugs. Honestly, who doesn't like to build a castle in the air to withdraw from everyday reality? If it works, the end justifies the means.

Of course we cannot present a complete dictionary of aphrodisiacs. There are several volumes of this kind listed in the Further Reading section. For the purpose of this book we just want to make a little excursion into this area and have a closer look at some of the better-known examples. As we will see, celebrity is by no means an indicator of effectiveness.

1.1   E is for Eggs

The Norwegian food scientist Bjödne Eskeland, based at the Matfork Norwegian Food Research Institute at Oslo has developed and patented a new method to produce a libido enhancing extract from chicken eggs, which he reckons will stir things up in the bedrooms. His studies of freshly fertilised chicken eggs show that these contain a particularly high proportion of amino acids and other vital substances after nine days. Double blind studies are said to have shown that the extract boosts vitality and sexual desire in men and women.

The extract, which is now commercially available under the name of "libido," is produced as follows: At the Norwegian chicken farm "Nesset Honseri," some 20,000 hens each lay an egg per day, on average; there are 2000 cocks to take care of fertilisation. Breeding machines incubate the eggs at 37 degrees Celsius and a controlled humidity of 85%, and turn them every two hours (as the hen would normally do, too).

After five to six days, the fertilised eggs are identified by means of light testing. On the ninth day, the egg white is taken from the fertilised eggs, freeze dried, pulverised and filled into capsules.

The pharmacologist Jürgen Reimann from Munich comments: "I have seen the studies of this preparation which is thought to enhance the libido, and judged by scientific criteria they are sound and plausible." He suspects that the effect may be due to endorphin fragments involving the amino acids leucine, glycine and tyrosine, in combination with antidepressive compounds like phenylamine and methionine as well as vasodilatory substances such as arginine.

1.2   G is for Gelée Royal and Ginseng

Another booster is gelée royal or royal jelly, the food that the worker bees produce for the queen bee, which is rich in vitamins and trace elements. The effects of this liquor on the development of a bee are striking. The selected larvae that receive it will develop into sexually mature queen bees within only 16 days. All the others will develop into the much smaller and sexless worker bees within 21 days.

From the perspective of food chemistry, gelée royal is rich in vitamins of the B complex, proteins, amino acids, hormone-like substances, minerals, and various trace elements. Medically, it is often used against the general ailments of ageing, but sometimes also as a cure of frigidity or impotence. Whether it has any of these effects remains questionable.

Figure 26 *Ginseng*
©Photodisk

The situation is similar for ginseng, the extract of a Korean root often offered in erotica shops as a "strengthening" tonic to increase the libido (Figure 26). From Korea, ginseng found its way into traditional Chinese medicine, was appreciated as an aphrodisiac in ancient China, and is now the most widely known medical herb from Asia. However, there have been very few studies that addressed the question of how well founded the claim of "medical" effects is.

In 2002, Bumsik Hong and his colleagues at the University of Ulsan, Korea, together with researchers from the Korea Ginseng and Tobacco Research Institute at Seoul conducted a study of 45 men with erectile dysfunction, to check whether ginseng could be considered a herbal alternative to viagra.

The participants took either 900 mg ginseng or a placebo three times a day for eight weeks. Following a two-week break, the groups were reversed, so the former placebo group received ginseng and vice versa, but neither the researchers nor the participants knew which was which.

Surprisingly, this study reported ginseng to have effects that were almost as significant as those of modern virility drugs including viagra (see next chapter). Around 60 percent of the participants experienced an improvement of their ability to achieve an erection.

It remains unclear, which mechanisms triggered by the plant might be responsible for the alleviation of erectile dysfunction. The Koreans speculate that ginseng increases the production of nitric oxide (NO), which dilates the blood vessels.

However, the US urologist Franklin C. Lowe from the St. Luke's Roosevelt Hospitel in New York casts doubts on this study. Apart from the relatively small number of participants, he criticises that the study did not involve men with severe virility problems, which might occur after surgical removal of the prostate, for example. Moreover, the precise composition of the preparation used in the study – so-called red ginseng produced from white ginseng by steam extraction – has remained unclear.

There is, however another ginseng preparation with a well-known composition. The Meskwaki Indians from the area that is now Iowa used to mix the local ginseng variant with mica soil, snake meat, gelatine and wild columbine. From this mixture, they brewed a proper love potion, which reportedly worked very well.

1.3   L is for Licorice

Licorice (*Glycyrrhiza glabra*) is a shrub commonly found in Southern Europe, Southwest Asia, and India. The strong sweet taste reminiscent of the eponymous sweets that are made from its extracts is mainly due to the content of glycyrrhizin, a sweetener 50 times more potent than common sugar. The plant has anti-inflammatory properties, and it is also a decongestant. For these reasons, it is commonly used in cough medicine.

So far, so good – but licorice wouldn't belong in the ABC of aphrodisiacs, if there wasn't an ancient belief in its powers as a love elixir. In India, especially, this belief has survived to the present day. A user's instruction from 1911 reads: "If you suck licorice together with the same amount of molten butter and honey, and drink milk with it, you achieve the maximal force for the execution of the coitus."

In Nepal, too, a tonic based on licorice is sold as an aphrodisiac to this day. The situation is similar in China, where licorice is found in every drug cabinet as a "depoisoning agent." In Germany, the association of licorice with love has survived in the expression "Süßholz raspeln" (literally: to grate licorice), for flirting. Nevertheless, we reckon that the presumed physiological effects of this harmless love potion is mostly based on the placebo effect.

1.4   O is for Oysters

Oysters remind me (R.F.) of a conversation which I had at Paris, back in 2000. I was invited by a French biotech company to join a briefing, and the company's PR man for the German market invited me for dinner at a typically French restaurant, where the cuisine turned out to be above all criticism. Not even the fact that the chairs appeared to be placed in the highest possible density could endanger the success of the evening.

After champagne for aperitif, we were served fresh oysters from Bretagne, and the conversation promptly turned to aphrodisiacs. My host remarked that this remedy was surely more wholesome than viagra. I replied that I had never experienced the slightest effect from the consumption of oysters.

He insisted: "In that case it's high time you increase the dosis." Fortunately, the two young French ladies, between whom I was almost squeezed into my seat, didn't catch up our conversation in German and showed no reaction.

Similar conversations must have taken place even 2000 years ago. The Roman poet Juvenal accused the harmless animal of indecency, saying it was a food suitable only for shameless and lascivious women. Among those who held the opposite opinion was Giacomo Girolamo Casanova, doctor of jurisprudence and knight of the Golden Spur. The famous lover was infamous for – among other things – his orgies fuelled by oysters and champagne.

China and Japan also have the traditional wisdom attributing miraculous effects to the Asian oyster *Crassostrea gigas*. The Danes have possibly the most explicit connection between the mollusk and sex, as their traditional name for the animal is "kudefisk," which literally means vulva fish.

However, as you may have suspected already, the oyster does not have much to offer by way of attractants or other active ingredients. Most probably, its semi-liquidity, its somewhat evocative appearance, and the taste of salt and sea are responsible for the psychological effect. The suggestion of luxury associated with a dinner of oysters by candlelight could further reinforce the placebo effect (Figure 27).

The situation is similar for caviar, which has also been attributed certain effects. Traditional wisdom holds that the sturgeon spawn is associated with a limitless natural fertility. Apparently, this transfers to the gourmets who crack the minuscule eggs against their palates. Its taste, symbolism, and association with luxury suggest why caviar, too, acquired the reputation of an aphrodisiac.

1.5   P is for Papaverine Injections

Before the discovery of viagra and similar drugs, which we will discuss in the next chapter, the most effective way of treating erection problems was by penile injection therapy. This method involved injecting one

Figure 27 *A dinner of oysters*
©Photodisk

millilitre of vasoactive substance through a small needle directly into the erectile tissue. Ten minutes after the injection, an erection will begin and last for an average duration of 80 minutes.

The Parisian surgeon Ronald Virag and the English physician Giles Brindley first discovered the beneficial effects that vasodilatory substances have on the erection. During a medical conference in Las Vegas in 1983, Brindley demonstrated the effect of such a treatment by injecting himself live on stage, and a few minutes later he is said to have opened his labcoat to present the result to a surprised audience.

Substances used in such injection therapies include papaverine, which is one of the natural ingredients of opium. This alkaloid was discovered in 1848 by Georg Merck and his coworkers at the chemical company E. Merck at Darmstadt, Germany. Compared to morphine, papaverine only has very little psychoactive effect. Instead, it is very effective in dilating the blood vessels, which is useful for the treatment of impotence.

Medical statistics show that the success rate of injection therapy – at some 96% – is very high, and significantly above that of viagra and similar products, which work their miracles only for about 70–80% of the users. The higher success rate suggests that injections can reach those patients that viagra cannot, *e.g.* when erectile dysfunction has an organic rather than a psychological cause. Thus, viagra and similar drugs may not be able to replace injection in the more severe cases.

As with every drug, papaverine has a few counterindications. Thus, patients who suffer from irregular heart rhythms or increased brain pressure will have to forego the therapy because of the associated risks. Among the unwanted side effects which can occasionally occur is the possibility of a painful, permanent erection that can last up to 36 hours, a phenomenon known as priapism. If the erection does not subsede after four hours, one should attend a hospital, where doctors can either inject an antagonist or remove the blood trapped in the erectile tissue. If this treatment is not applied in time, the tissue can be damaged. Most doctors see priapism as a minor risk. While there are no precise figures, the rate is said to be significantly below one percent.

Papaverine preparations are available on prescription, typically as papaverine hydrochloride. Doctors recommend not to use them more often than twice a week, to avoid inflammation and scarring of the skin of the erectile body.

1.6   R is for Rhinoceros Horn

The crushed horn of the rhinoceros is one of the oldest, most famous, and also most controversial aphrodisiacs. The fascination with this animal can be traced back to stone age hunters and seems to have survived in today's pharmacies in China.

Thus, many people in the Far East are convinced that the rhinoceros's horn possesses unique powers. In the 16th century AD, the famous Chinese pharmacist Li Shih Chen (1518–1593) wrote his comprehensive opus "Pen ts'ao Kang-Mu." With its 50 volumes it contains no less than 12,000 recipes and is thought to be the most detailed Chinese collection of pharmacological information. This work cites as the main applications of the rhino's horn (xi jiao in Chinese): snake bites, obsession, hallucination typhoid fever, headaches, furuncles, burns, fever, and of course impotence. If the horn was burned and mixed with water, it could also serve to cure ptomaine poisoning and vomiting.

Li Shih Chen warned emphatically that the preparations should not be given to pregnant women, as they might lead to miscarriages. Most of the "xi jiao" preparations found in Chinese pharmacies today have still been prepared strictly by the rules set by Li Shih Chen.

But which compounds are found in the legendary horn of the rhino? Scientific analysis shows that it consists of the proteins eukeratine and keratine, exactly like human hair and fingernails. Further ingredients include peptides, amino acids, guanidine derivatives, cholesterol, calcium carbonate and phosphate. All in all, these substances are not going to boost anybody's virility in any way.

1.7 S is for Spanish Fly

"Insatiable sex drive. Tingling in your sex organs. Tendency to let yourself go sexually. Original Spanish Fly D5 can be mixed with any drink. For men and women."

Well, this is at least what a certain mail-order company promises its clients, obviously unaware of the problems that can arise from substances that "can be mixed with any drink," especially if someone mixes them into a drink that isn't his own. However, as the abbreviation D5 points to homeopathic dilution, the concoctions offered as "Spanish fly" should be completely harmless. Misuse aside, any non-homeopathic doses, would be worrying for toxicity reasons, as we shall see below.

In fact, the animal from which the "Spanish fly" concoction is derived isn't a fly at all, but a metallic green shimmering blister beetle of the genus *Cantharis*, which is found mainly in Southern Europe, but also in Asia, with a related genus in South America. In their dried state, these "Spanish flies" contain around one percent of the highly toxic compound cantharidin, which can be extracted with ether. Just 30 milligrams of this relatively simple organic compound of just 10 carbon atoms can be fatal for an adult. Accordingly, the toxicological literature reports many deaths from this poison. This is why you will not find any pharmacologically active preparations of this substance in any pharmacy, but only the homeopathic ones (Figure 28).

The Spanish fly owes much of its fame in Europe to the notorious Marquis de Sade (1740–1814). In a historic account of one of his excesses we read: "It is reported from Marseille that the Marquis de Sade, who in 1768 created such a stir with his crimes against a whore, with whom he allegedly wanted to try a new cure, has just produced an initially amusing but in its later consequences horrible spectacle. He gave a ball to which he had invited many people, and for dessert he distributed very nice chocolate pastils, of which many people ate.

Figure 28 *Spanish fly*

However, these had been prepared with pulverised Spanish fly ... All who had eaten them were possessed by a shameless lust and committed the wildest excesses of love. The feast degenerated to a wild ancient Roman orgy. The most chaste women could no longer withstand their lust. The Marquis de Sade abused his sister in law, with whom he then escaped ... several people died from the consequences of the excesses, others are still very sick ..."

If cantharidin is applied to the skin it causes inflammation with blisters. This effect has been known since ancient times. The Egyptian papyrus known as the Ebers papyrus, the oldest surviving book on medical science, describes a catharidin plaster. It was applied to the belly of women giving birth, allegedly "to help the child detach." We can imagine that this ordeal must have sped up labour.

The Inkas used the blood of a related kind of beetle to remove corns and warts. Aristotle, Hippocrates, Pliny, and Galenus all reported both a limited therapeutic effect and the toxicity of the "Spanish fly."

The aphrodisiac effect of the beetle concoction has also been known since ancient times. A report from 1572 states: "A woman said under oath that her husband, after taking the drug, had intercourse with her 40 times in one night. Even during his interrogation he was still so excited from the drug that he implored the examiners to let him die in his ecstasy, which appears to have happened according to his wishes soon afterwards."

As mentioned above, the Marquis de Sade brought about a renaissance of the "Spanish fly" and ensured its lasting popularity, which apparently survives to this day.

1.8   T is for Tiririca

Many gardeners appreciate this plant which is originally from East Africa and Madagascar. Tiririca (*Cyperus rotundus*), a relative of the Egyptian paper reed from which papyrus is made, is a sedge with inconspicuous flowers located at the top of the stem, shielded by an umbel of bracts. This perennial has a reputation of being robust, undemanding, and very easy to keep.

What many gardeners may not know about this plant is that it is one of the classical aphrodisiacs, even though its effectiveness remains unclear. Nevertheless, the Egyptians appreciate the tubers of the plant as an aphrodisiac to this day. Similarly, some ethnic groups in East Africa like to squash the tubers with milk and use the result as a love potion. As there are also reports from China and the Amazon region supporting a stimulating effect of the plant, it has to be asked whether this is one of the few "real" aphrodisiacs.

The only fact that scientists have so far been able to establish in response to this question is a fairly high content in essential oils. Maybe the enticing fragrance that spreads during preparation of the tubers is responsible for the stimulating effect.

1.9 V is for Vitamin E

From around 1970 through to the mid 1980s, a miracle drug known as vitamin E haunted the media. Many broadcasts and magazine articles attributed fantastic effects to the vitamin, chemically known as alpha-tocopherol. Even today it is sometimes referred to as a virility or fertility vitamin. But does this description have any foundation in science?

What is clear is that vitamin E has been known since 1968 as an essential nutrient. Normally, we ingest sufficient amounts of this vitamin with our daily meals, as it is contained in many kinds of food. Wheat germs and all kinds of plant oil contain large amounts of the vitamin, as do butter, nuts, various vegetables, and soy beans.

Presumably, animal experiments were responsible for the exaggerated reactions and expectations. On several occasions, researchers had found that sterile laboratory rats were able to conceive after administration of vitamin E. However, reports tended not to mention the fact that similar experiments with other species didn't show any success. With human subjects, no effect on reproduction could be found.

In summary, vitamin E is like other vitamins, good for your health and general strength. Forcing it into the role of an aphrodisiac, however, would be audacious.

DOES ALCOHOL HAVE A STIMULATING EFFECT ON WOMEN?

A study published in *Nature* claims to have shown that the testosterone production in women – and thus their sex drive – is increased after consumption of alcohol. This contrasts with the observation in men that alcohol removes inhibition but at the same time weakens virility.

At the end of the day, it probably comes down to the right dosis, as Paracelsus recognised already. Thus, the study concludes that both men and women should not consume more than half a gram of alcohol per kilogram body weight prior to sexual encounters, if they don't want to suffer from a loss of performance. For a person of 75 kg, this critical dosis would be reached roughly with half a bottle of wine.

1.10   Y is for Yohimbine

After all these more or less dubious candidates, you may ask whether there is no reliable aphrodisiac in the treasure trove of Nature. And mankind's old companion, alcohol, doesn't do much good, either (see Box on p. 151). Well, there is one at least, namely yohimbine. It is produced from yohimbe, the bark of the yohimbe tree (Figure 29).

The evergreen tree, reminiscent of an oak, grows in the tropical forests of Cameroon and Nigeria and can reach heights of up to 30 metres. In Africa, extracts from its bark have been used as an aphrodisiac for centuries. Sailors who reported the miraculous effect of the bark at the end of the 19th century brought the knowledge to Europe.

For the preparation of yohimbine, the bark must be dried and crushed. The resulting granulate can be extracted with alcohol and taken as a tonic, but it can also be used to prepare a kind of tea.

Experts suspect that yohimbine – chemically speaking an indole alkaloid – can block a specific class of receptors, known as the α_2-receptors. These control the release of the hormone noradrenaline, so when they are blocked, there will be more noradrenaline released into the body.

The excess of noradrenaline increases the responsiveness and sexual readiness. Moreover, scientists suspect, it dilates the arteries in the genital region and thus improves blood flow. Lastly, it may also increase the blood levels of two other hormones, the happiness hormone serotonin, and the drive-enhancing dopamine.

A little bitterness to spoil the good news: Yohimbine is not suitable for immediate action. All studies have shown that the drug has to be taken regularly over a duration of two to three weeks before it begins to show a significant effect. This, too, suggests a complex interplay with the hormones involved. Thus, in contrast to other virility drugs, yohimbine is no "one-night-stand." It is rather a drug that has to be taken over a long period, like hormone therapy.

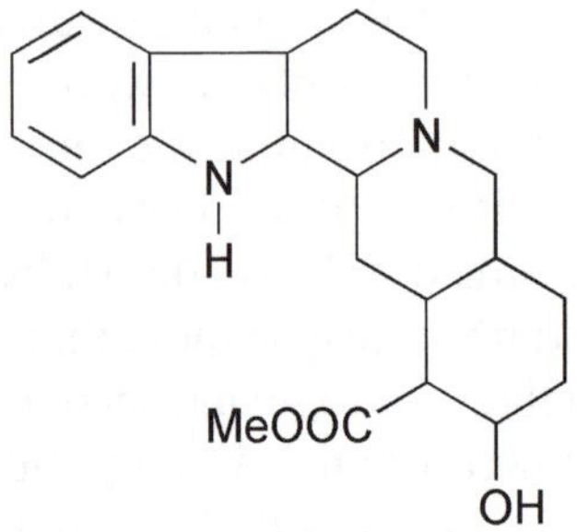

Figure 29   *Yohimbine*

CHAPTER 18

Viagra & Co

"Love me a little less, but longer."
Jewish proverb

1   A GROWING PROBLEM

For Bianca and Michael this is not (yet) a problem – as a young couple they don't need virility drugs. In her studies, however, Bianca has heard that virility problems are relatively frequent in middle-aged men, and that people start talking of an "epidemic."

Many studies have shown that the loss of virility phenomena collectively known as erectile dysfunction (ED) is on the rise even on a global scale. For example, in the "Massachusetts Male Aging Study," the data of 1290 men aged between 40 and 70 years were analysed.

The results speak for themselves. Overall, 52% of the participants reported virility problems of some kind or the other, with the prevalence of ED naturally increasing with age. For the 70-year-olds, the ED rate was 67%, but even among the 40-year-olds, 38% said they frequently had such problems.

It is remarkable that the topic of ED is still a taboo, obviously because it doesn't fit in with common conceptions of being male. Thus, international estimates suggest that – regardless of the loss in quality of life – only one in ten men with the disorder seek treatment. Obviously, most men are unaware that ED responds well to therapy.

Uwe Hartmann at the Medical University of Hannover, Germany, diagnoses a proper "speechlessness" of male patients: "Overall, erection problems constitute an underdiagnosed and untreated disorder, which undermines a man's ego, reduces his quality of life, can lead to psychosomatic problems, depression or alcoholism, and often ends with chronical disorders and in loneliness," he summarises.

Among the traditional possibilities for the treatment of erection problems, the injection with erection-enhancing drugs such as papaverine (see previous chapter) has become the most popular one. Since 1983, it has been used to great success in the treatment of ED, but its key disadvantage is the requirement of an injection. Which man – apart from fans of extreme piercing – will relish the idea of stabbing their penis with a needle before they can proceed to have sex?

1.1   The Discovery of Viagra

What do teflon and viagra have in common? Honi soit qui mal y pense – if you were expecting a slippery joke, you will be disappointed, as we are entirely serious about this matter. The answer is that both of these highly successful products were discovered by sheer serendipity. Teflon turned up as a white powder following an explosion in a laboratory. For viagra, it was the unexpected side effect of a drug tested against heart disease that stole the lime light. When the US company Pfizer conducted clinical studies of their substance "sildenafil" with selected patients suffering from heart disease, the male participants reported long lasting erections, and were so pleased with this side effect that they didn't want to stop using the pills. Sheer good luck had produced a new blockbuster drug for the pharma giant.

As viagra can be taken as a pill, it is not just a potent but also a convenient treatment for erectile dysfunction. Analysts have estimated that there are around 140 million potential customers for it around the world. However, contrary to a widespread misconception, viagra is no aphrodisiac, as it doesn't enhance sexual desire. As a urologist put it: "If you sit calmly on your chair and wait for an effect to show up, you will be disappointed." So what exactly is the fuss around the little blue pill about?

In 70 to 80% of men, viagra works very well, but only as a secondary support. Only when a man gets into the sexual mood, the pill will ensure a better and lasting erection. But how does this uplifting effect come about? To answer this question, we will have to take a closer look at the little blue diamond, as the drug is affectionately called by some of its fans.

1.2   The Penis as a Book-Keeper

To the chemist, viagra is a phosphodiesterase inhibitor. It achieves its positive effect on the male erection by interfering with a complex interplay of enzymes, which we will now have to unravel.

In principle, an erection is based on a closed control loop, namely the balance of blood flow into and from the erectile tissue of the penis. The influx is controlled by ring-shaped muscles in the walls of the organ's artery. In the absence of an erection, these muscles are strained, which closes the vessels.

Following sexual stimulation, the messenger cGMP (cyclic guanosine monophosphate) is produced in the muscle fibres of the artery walls. This messenger makes the muscles relax, so the artery allows the blood to flow in faster than it can flow out, which leads to an accumulation of blood, *i.e.* an erection. Thus, the penis is nothing but a book-keeper.

However, matters are made more complicated by the enzyme phosphodiesterase type 5 (PDE-5), which seeks to suppress the effect of cGMP by chemically cleaving the messenger molecule. As a love killer, PDE-5 thus acts against the erection. As you will have figured out by now, this is exactly the point where viagra exerts its effect. The compound sildenafil blocks this enzyme and ensures that even relatively minor amounts of cGMP, which are still found even in men with a pronounced ED, can lead to a complete relaxation of the artery muscles and thus to a lasting erection.

1.3 Viagra – Just What the Doctor Ordered?

Most medical experts have a generally positive opinion of viagra, but they also warn of potential risks. For example, the urologist Hartmut Porst from Hamburg, Germany, who has over 20 years of experience in this field, says that viagra is a very good drug, but that it should not be misused. For example, patients who have received a prescription for the drug should not pass it on to their friends. Nobody should take the drug without consulting their GP about possible interactions with other drugs.

Another urologist hits a similar note in warning men, not to use viagra as a "party drug." "Viagra is a therapy against male impotence, it is a drug and by no means a virility booster," the expert clarifies.

"Viagra doesn't convert you into a 21-year-old, but it actually solves the problem," summarises Robert Shay from Los Angeles. The 75-year-old researcher has taken part in the clinical trials which found a 70–80% success rate with only acceptable side effects. According to TIME magazine, there is a one in ten risk of severe headaches, and three percent of patients experience temporary vision impairment.

It is clear that the use of viagra leads to a general widening of blood vessels. In a healthy person, this manifests itself in a reduction of blood pressure by around 10 mm Hg. If another vessel dilating drug is taken at the same time, the decreases in blood pressure can add up to a dangerous extent. This is a particular risk factor for nitrates, a group of substances often used by men of advanced age for the treatment of various heart problems, such as angina pectoris.

1.4   Potent Drugs Fight Impotence

By now, viagra is facing several competitors. Bayer AG, for instance, has obtained licences for its compound vardenafil both in the US and in Europe. The substance, which Bayer and Glaxo-SmithKline are marketing under the name of levitra, is a PDE-5 inhibitor just like viagra. Its effectiveness was confirmed in a study involving 4,000 men aged 21 to 71, covering the entire spectrum of erectile dysfunction, including both purely organic and purely psychological problems, along with mixtures of the two. All of these were represented as serious and intermediate cases.

Hartmut Porst – whom we met above – oversaw this study. "The compound vardenafil convinced us, and it did so in all groups of patients. It is safe and well tolerated," he summarises. By the end of the three-month trial period, around 75% of all sexual encounters across all groups could be completed successfully. This is three times more than before treatment.

According to Bayer and Glaxo, specialists have welcomed the choice of therapy brought about by the introduction of levitra. "The informed patient can express his wishes as to which of the drugs should be prescribed," says one urologist. Another drug that is also based on a PDE-5 inhibitor is cialis (active compound: tadalafil), which has been introduced by the US company Lilly. For this drug, too, test results show a similar success rate as for the two competitors.

One thing should not be overlooked, however: For the majority of patients, it's not just the drugs, but also the right words which are crucial for the success of the therapy. "The therapy of the physical problem works very well, thanks to the new drugs," explains Ulrike Brandenburg, a specialist for psychotherapeutic medicine in Aachen, Germany. "Bigger problems often arise from the fact that the patient has to relearn sexuality and intimacy with his partner. This isn't easy in the beginning, and it may not be fun to start with!" Brandenburg's experience has shown that up to eight "attempts" may be needed before a couple has adapted to the change. "Therefore, the therapy should be discussed with the partner before the first use of the pill, or a joint appointment with the doctor should be made," she recommends.

CHAPTER 19

Epilogue: Returning from the Airport

"Any marriage, happy or unhappy,
is infinitely more interesting than any romance,
however passionate."
W.H. Auden (1907–1973)

1 THE IMPORTANT THING IS THAT THE CHEMISTRY WORKS – THE REST REMAINS A MIRACLE

"Where did you leave the car?" Bianca looks at Michael. "In P6, I couldn't find anything better," replies Michael, who guides the trolley with the luggage in the direction of the carpark. "Back home I have prepared a little something to eat."

Bianca smiles at him. "Oh, that sounds seductive, has my personal chef donned his apron after weeks of abstinence?" "No, but I got some sushis from the delicatessen, hope that's okay?"

"Absolutely," replies Bianca. "I must have gained three kilos. The American steaks were simply irresistible."

After a one hour drive, the two reach their flat. Michael carries her suitcases upstairs. While she begins to unpack, he lays the table with much care and puts on a CD with love songs. When Bianca enters the dining room, she notices the candles on the table.

"How did you spend the evenings and weekends?" Her eyes shine. "Hope not only at the computer, or did you?"

"No, no," replies Michael. "I have also read quite a bit. You know, normally I rarely find the time for that."

"And what did you read?"

Michael opens a drawer and hands her a present. "This one, for example."

Bianca removes the wrapping carefully and holds a book in her hands. "About the chemistry of love," she whispers. "What's that all about?"

"Well, these authors – a married couple, by the way – have looked closely at the human brain and comes to the conclusion that everything we think and feel is ultimately based on chemistry. Did you know that?"

"Mmmmhhh," Bianca replies, as her professional pride as a future medical doctor feels slightly tickled. "There is definitely some truth in that."

Michael looks deep into her eyes. "I sense a 'but' ..."

"I'm just thinking of the words of my professor," replies Bianca. "He always told us students how he didn't believe that the human brain reflecting upon itself could ever reveal all of its secrets and miracles."

Michael tenderly takes her hand and whispers into her ear: "As long as our chemistry works – the rest remains a miracle."

Further Reading

Allende, Isabel: Aphrodite, Perennial, London 2005

Angier, Natalie: "Pleasure Hormone Reveals New Facets: Oxytocin linked sexual, social well-being", New York Times, 22. 01. 1991.

Angier, Natalie: Woman – an intimate geography, Virago, London 1999.

Aron, Arthur et al.: Reward, motivation and emotion systems associated with early-stage intense romantic love, J. Neurophysiol. 2005, 94, 327–337.

Bickel, Margot; Spring, Anselm: Sehnsucht die verwandelt; Franckh-Kosmos, Stuttgart (1995).

Bradt, Steve: Pheromones in Male Perspiration Reduce Women's Tension, Alter Hormone Response that Regulates Menstrual Cycle; Penn News (University of Pennsylvania) 14. 03. 2003.

Brody, Howard; Brody, Daralyn: The placebo response – how you can release the body's inner pharmacy for better health, HarperCollins, London 2000.

Burnham, Terry; Phelan, Jay: Mean genes, Perseus, Cambridge (Massachusetts).

Buss, David: The evolution of desire: Strategies of human mating, Basic Books, New York 2003.

The Chimpanzee Sequencing and Analysis Consortium: Initial sequence of the chimpanzee genome and comparison with the human genome, Nature 437, 69–87 (2005)

Crick, Francis: Astonishing hypothesis – Scientific search for the soul, Pocket Books, New York 1995.

Crick, Francis; Koch, Christof: The problem of consciousness; Scientific American Special issue: The hidden mind, January 2002.

Damasio, Antonio R.:Looking for Spinoza, Vintage, New York 2004.

Damasio, Antonio R.: The feeling of what happens – body, emotion, and the making of consciousness, Vintage, New York 2000.

Damasio, Antonio R.: Descartes' error – emotion, reason and the human brain, Penguin, London 2005.

Dawkins, Richard: The selfish gene, 3rd edition, Oxford University Press, Oxford, 1989, 2006.

Djerassi, Carl: This man's pill – reflections on the 50th birthday of the pill, Oxford University Press, Oxford, 2003.

Egan, Kate: Love & Sex – The vole story; Emory Medicine, Emory University, Summer 1998.

Emanuele, Enzo et al., Raised plasma nerve growth factor levels associated with early-stage romantic love. Psychoneuroendocrinology 2005, (advance online publication end of November 2005) doi: 10.1016/j.psyneuen.2005.09.00

Emsley, John: The consumer's good chemical guide, WH Freeman, New York, 1994.

Emsley, John: Molecules at an exhibition, Oxford University Press, Oxford, 1999.

Eskeland, Bjödne et al.: Sexual desire in men: Effects of oral ingestion of a product derived from fertilized eggs. J. Int. Med. Res.1997, 25, 62–70.

Fisher, Helen: The first sex: the natural talents of women and how they are changing the world, Ballantine, New York 2000.

Fisher, Helen: Anatomy of love – a natural history of mating, marriage and why we stray, Batus, 1995.

Fisher, Helen et al.: Romantic love: An fMRI study of a neural mechanism for mate choice. J. Comp. Neurol. 2005, 493, 58–62.

Flynn, R. J.; Williams, G.: "Long-term Follow-up Patients with Erectile Dysfunction Commenced on Self-injection with Intracaversonal Papaverine with or without Phentolamine", British Journal of Urology, 78, 628–631 (1996).

Giles, Graham: BJU international 92, 211 (2003).

Goleman, Daniel: Emotional Intelligence, Bantam Books, New York 1997.

Gregory, Richard L., Eye and brain – the psychology of seeing, Oxford University Press, Oxford 1997.

Gross, Michael: Jacobson's molecules. Chemistry in Britain 38, No. 11, 20 (2002).

Gross, Michael: Cupid's chemistry. Chemistry World 3, No. 2, 32–35 (2006).

Güntürkün, Onur: Adult persistence of head-turning asymmetry. Nature, 421, 711 (2003).

Ivell, Richard: The fate of the male germ cell, Plenum Press, New York 1997.

Kilham, Chris: Hot plants. Griffin, Chicago (2004).

Mäckler, Andreas: Was ist Liebe..? 1001 Zitate geben 1001 Antworten; DuMont Buchverlag, Köln (1988).

Marazziti, Donatella; Canale, Domenico: Hormonal changes when falling in love. Psychoneuroendocrinology 29, 931 (2004).

Morris, Desmond: The naked woman – a study of the female body, Jonathan Cape, London 2005.

Neumann, Inga: Effects of Suckling on Hypothalamic-Pituitary-Adrenal Axis Responses to Psychosocial Stress in Postpartum Lactating Women; Journal of Clinical Endocrinology & Metabolism, Vol. 86, No. 10, p. 4798–4804.

Sternberg, Robert J.: Love is a story – a new theory of relationships, Oxford University Press, Oxford 1998.

Süskind, Patrick: Perfume – the story of a murderer, Penguin, London 1989.

Watson, James D.: The double helix, Penguin, London 1999

Watson, Lyall: Jacobson's organ and the remarkable nature of smell, W.W. Norton & Company, New York 2000.

Zinberg, Norman E.: Drug, Set and Setting – the Basis for Controlled Intoxicant Use, Yale University Press, New Haven und London 1984.

Subject Index

acetylcholine 27–28
Adam's apple 63–64
adrenaline 49–51, 56
adrenocorticotrophic hormone (ACTH) 56
aglandular hormones 51
Alcmaeon of Croton 3
alcohol 151
aldehydes 124
alkaloids 114
ambergris 125
amino acids 43, 44
Anaxagoras 3
androgen hormones 56
androstenol 135
androstenone 135
antagonist action 34
aphasia 17
aphrodisiacs 142–143
  alcohol 151
  chicken eggs 143
  gelée royal (royal jelly) 143–150
  ginseng 144–145
  licorice 145–146
  oysters 145–146
  papaverine 145–146
  rhinoceros horn 148–149
  spanish fly 149–150
  tiririca 150–151
  vitamin E (tocopherol) 151
  yohimbine 152
apical lobes of the brain 9
Aplysia (sea slug) 20
arguments, domestic 67
artificial pheromones 138–139

atomic fusion 43
attractiveness of the opposite sex 36–37, 39–40 117, 128–129
autism 77
autonomous (vegetative) nervous system 33–34
axons 18

babies
  and endorphins 111–112
  role of adrenaline at birth 51
basal temperature 62
bipolar disorder 53
birth control 87–91
black cohosh 96
blood 25–28
blood pressure, high 48
Blothner, Dirk 38–41
brain
  ageing and loss of nerve cells 7
  architecture 6–7
    brain stem 7
    cerebellum 8
    cerebrum 7, 10
    cortical lobes 9–11
    difference between the two halves 10, 13–15, 16–17
    hypothalamus 7
    limbic system 8–10
    spinal cord 11–12
  direct communication with computers 22–23
  early ideas 2–5
  energy usage 30–31
  glucose usage 31–32

learning 13
mapping 5–6
measuring thoughts 17–19
MRI scanning 16–18, 25
parallel processing of information
12–14
signalling 27–30
tumours 32
breastfeeding
endorphins 111–118
oxytocin 72
Brindley, Giles 146
Broca, Paul 5
Broca region of the brain 17
Brodman, Korbinian 5
Brody, Howard 107–114
Buss, David 39–40
Butenandt, Adolf 89
butterflies in the tummy 2

calcitonin 52–53, 56–57
cancer, inherited tendencies 48
cantharidin 149, 150
Carmichael, Mary 75–76
Casanova, Giacomo Girolamo 146
castor (scent from beavers) 125
cauda equina 12
cells, chemistry of 19–20
central nervous system (CNS) 11
cerebellum 8
cerebrum 7, 10
chemical elements 44–47
childbirth
role of adrenaline 51
role of endorphins 111
Chinese medicine 108
chocolate 83, 85
chromosomes 47–48
cibet 125
Cleopatra 121–122
clitoris, response to stimulus 35
clothing, attractiveness to men 40, 128–
129
cohoch, black 96
colour blindness 48
colours 128–135
communications between computers and
the brain 22–24
contraception *see* birth control

Contzen, Barbara 38–39
Coolidge, Calvin 80
Coolidge effect 80–81
copulation *see* sex
copulins 137–139
corpus luteum 61
cortical lobes 9–11
cortisol 57
cosmetics 118–119
Coty, Francois 124–125
Crick, Francis 45–46
crocodile excrement, as a method for birth
control 88
cuddling
after orgasm 69
dealing with pain 110–117
cyclic guanosine monophosphate (cGMP)
21, 22, 133, 155

dark current 21
dehydroepiandrosterone (DHEA) 57,
98–99
dendrites 18
Descartes, René 5
deoxyribonucleic acid (DNA) 43–48
divorce rates 86
Djerassi, Carl 89–90, 92–94
Doermer-Tramitz, Christiane 36
domestic disputes 67
dopamine 28, 57, 78–79, 152
addictive nature 79
Coolidge effect 80–81
Dorey, Grace 97
double-helix of DNA 45–46

Earth, origin of 42–43
Egarter, Christian 95
egg cells
attraction of sperms 132–133
development of 61
eggs, chicken, aphrodisiac properties 143
Egypt, Ancient 3, 87–88
electrical impulses 18–20
electrocardiograms (ECGs) 25
electroencephalograms (EEGs) 25
electrooculograms (EOGs) 25
emotions
effects of phenylethylamine 85–86
hormones 25–27

location in the brain 9
 neurotransmitters 28–29
endogenous opiates *see* opiates,
 endogenous
endomorphines 104–105
endorphins 57, 102, 105–115
Erasistratos 3
erectile dysfunction (ED) 153–154
erection of penis 35
 problems with 97–98, 147, 153–158
Eskeland, Bjödne 143
estradiol 57, 61
estrogen hormones 57, 61
 onset of menopause 94
evolution of species 42–44
eye contact 37–38

Farina, Giovanni Maria 122
fentanyls 115, 116
Fink, Gereon 16–18
Fiorine, Dennis 79, 80–81
Fischer, Ernst Peter 45
Fisher, Helen 68–69, 71–72, 86
flirting 37, 38–40
follicle stimulating hormone (FSH) 57
 action in men 63
 action in women 57
follitropin 57
Francke, Wittko 132, 135, 139
frontal lobe of the brain 10
Fuchs, Bernhard 22–23
functional imaging in real time (FIRE)
 MRI 27
functional magnetic resonance imaging
 (fMRI) 16–18, 26

Galenus, Claudius 3
Gall, Franz Josef 5
gamma-aminobutyric acid (GABA) 27,
 28, 30
glandular hormones 51
gelée royal (royal jelly), aphrodisiac
 properties 143–144
genes 44, 45–48
genetic code 47
genetic influences 41–42, 45–46
 inherited disorders 48
 pheromones 139–140
Giles, Graham 65–66

ginseng, aphrodisiac properties 144–145
Glaser, Ronald 67
glucagon 54
glucose
 metabolism 54
 radiolabelling 32
 usage by the brain 31–32
glutamic acid 22, 29
glutamic acid rich protein (GARP) 22
glycine 28
glycyrrhizin 145
Godfrey, Peter 86
Goldstein, Avram 113
Grammer, Karl 36, 135–137
growth hormone (somatotropin) 59–60
Güntürkün, Onur 15–17

Hal (computer in *2001 – A Space Odyssey*)
 22–23
Hartmann, Uwe 153–154
Hatt, Hans 133
heart, pulse rate 33
hemispheres of the brain 7
 difference between 10, 14–15
 during kissing 15–17
hemoglobin 26
heroin 114, 115
Herophilus 3
highs from sports 106
Hippocrates 2, 88
Hong, Bumsik 144–145
hormone deficiencies 51–53
hormone replacement therapy (HRT)
 94–95
hormones 7, 49–60
 emotions 25–27
Hubble space telescope 44
Hughes, John 102
human leukocyte antigens (HLA)
 140–141
Hummel, Thomas 134
hypothalamus 7, 10, 61

immune system 67
inheritance of genes and characteristics
 47–48
Insel, Thomas 76–77
insulin 54, 57
intercourse *see* sex

internal clock 54
ion channels of cells 19–22
Ivell, Richard 68, 73–74, 75, 76
Iverson, Leslie 30

Jacobson, Ludwig Levin 134
Jahnke, Volker 134–135
Jentsch, Thomas 20
Jung, C.G. 8

Kaupp, Benjamin 21–23, 132–133
kissing, left-and-right bias 15–17
Knowlton, Charles 89
Kobal, Gerd 134
Kosterlitz, Hans 102
Kubrick, Stanley 22

Lancet, Doron 126
language understanding, location in the
    brain 10, 17–18
larynx, development at puberty in males
    63–67
learning 13
left-handedness 10
    kissing 15–17
Leonardo da Vinci 4, 5
letters of the genetic code 46, 47
levitra 156
Li Shih Chen 148–149
libido (sex drive) 26, 54
licorice, aphrodisiac properties 145–146
limbic system 9–10
lipsticks 117–119
lock-and-key mechanism 103–104
love
    ancient explanations for 60
    at first sight 36–37
    mood changes 55
    romantic 69
Lowe, Franklin C. 145
LSD 28
luteinising hormone (LH) 57, 137
    action in men 63
    action in women 61

magnetic resonance imaging (MRI) 25
magnetoencephalograms (MEGs) 25
major histocompatibility complex (MHC)
    140

mammalian brain 8–10
Marazziti, Donatella 78–79, 84–85
masturbation 65–66
McClintock, Martha 139
mediator hormones 56
melatonin 58
memories 20
men
    attractive characteristics of women 40,
        117, 128–129
    attractiveness of pheromones 137–139
    characteristics developed at puberty
        63–64
    contraceptive pill 92
    effects of oxytocin 72
    effects of prolactin 112
    erection problems 97–98, 147,
        153–158
    eye contact 37–38
    flirting 38–39
    'male menopause' 96–105
    penis 154–155
    pheromone production 136–137
    tolerance of women's behaviour
        75–76
menopause 93–94
    hormone replacement therapy (HRT)
        94–95
    'male menopause' 96–99
    symptoms 93–94
    treatments from nature 95–96
menstruation 61
    follicle stimulating hormone (FSH) 57,
        61
Merker, Hans-Joachim 135
microparanoia 84–85
milk production 62
Miller, Stanley J. 43
Monroe, Marilyn 124, 128
mood changes 52–53
    first love 55
morphine 104–105, 114
movement control, location in the brain 9
musicality, location in the brain 10
musk 125

Napoleon Bonaparte 122, 137
naxolone 112–113
near death experiences 110

Neave, Nick 136–137
nerve cells 6, 12
  axons 18
  dendrites 18
  loss on ageing 7
  synchronisation of neuronal activity 24
nerve networks 35–38
neuropeptides 27
neurosecretory hormones 51
neurotransmitters 27–30
nitric oxide (NO) 145
noradrenaline 28–29, 50, 58, 152
norethindrone 90
nucleic acids 43–44

Odent, Michel 51, 110–112
opiates, endogenous 100
  discovery 101–102
  effects of complete absence 113
  endorphins 105–106
  following birth 111
  mode of action 103–104
  natural 114
  naxolone 112–113
  near death experiences 110
  pain research 109–110
  placebo effect 107–108
  receptors 102–103, 116
opioids, synthetic 114–115
  anti-terrorist measures 115–116
opium 114
optical illusions 15
orchids 131
orgasm, cuddling afterwards 69
origins of life 42–44
Ornstein, Robert 18, 104, 103–104, 113
oxygen
  content of the blood 26
  usage by the brain 30
oxytocin 58, 68–69
  breast feeding 72
  coping with stress 72–73
  cuddling after orgasm 69
  effects on men and women 72
  partnership and fidelity 76–77
  Pavlovian response 70–71
  social behaviour 76
  variety of effects 73–74
oysters, aphrodisiac properties 145–146

Pääbo, Svante 126–127
pain
  endorphins 108–109
  hugging 110–111
papaverine, aphrodisiac properties
  145–146
parasympathetic nervous system 34
parietal lobe of the brain 10
Parkinson's disease 24, 78, 86
Pavlov, Ivan Petrovich 70–71
Pearl index 90
penis 154–155
  erection 35
  erection problems 97–98, 147, 153–154
perfumes *see* scents
personality types 132
Pert, Candice 101–102
Pfaff, Donald Wells 107
phenylethylamine (PEA) 58, 85–86
pheromones 130–131
  see also scents
  artificial 138–139
  differences between men and women
    135–138
  evolutionary aspects 134–135
  genetic influences 139–140
  production 135
  role in humans 131–132
Pincus, Gregory 89–90
pituitary gland 7, 10
placebo effect 107–108
planets, origin of 42–43
pneuma 5
pomegranates, contraceptive properties
  87–88
Pöppel, Ernst 12–14
Porst, Hartmut 98–99, 155, 156
positron emission tomography (PET)
  scans 31, 32
  hormone deficiencies 54
pregnancy 62
priapism 148
processing of information by the brain
  12–18
progesterone 58–59, 61, 62
prolactin 58, 112
prostate cancer 65–69
proteins 43
pseudogenes 129

psi phenomena 18
psilocybins 28
puberty 62
pulse rate 33

reflexes 12
Reimann, Jürgen 143
Rendell's 89
reptilian brain 8
reward system 30
Rhases 88
rhinoceros horn, aphrodisiac properties
      148–149
rhodopsin (visual pigment) 21
ribonucleic acid (RNA) 44, 47–48
right-handedness 10
    kissing 15–17
romantic love 69
royal jelly (gelée royal), aphrodisiac
      properties 143–144

Sade, Marquis de 149–150
scents 120
    see also pheromones
    animal origins 125
    colour vision 128
    historical aspects 120–123
    modern formulations 124–125
    notes 123–124
    sense of smell 125–126
Schäfer, Michael 108
schizophrenia 77
Schröder, Dietmar 22–23
sea slug (*Aplysia*) 20
seasonal affective disorder (SAD) 83
serotonin 28, 59, 82–85
sex
    beneficial effects 66–67
    endorphins 110–111
sex drive (libido) 26, 54
Shay, Robert 155
sildenafil *see* viagra
sleep, neurotransmitters 28–30
smell, human sense of 125–126
Smith, Timothy D. 128–129
sneezing, as a method for birth control 88
Snyder, Solomon 101–102
sodium ions
    nerve communication 18

vision 21–22
Solar System, origin of 42–45
somatic nervous system 33
somatotropin (growth hormone) 59–60,
      98
Soranus of Ephesus 88
soybeans, use during menopause 96
spanish fly, aphrodisiac properties
      149–150
Spehr, Marc 133
spermatozoa
    attraction to egg cells 132–133
    follicle stimulating hormone (FSH) 57
    testosterone 63
spinal cord 11–12
St John's wort, use during menopause 96
Stangl, Werner 6, 7, 15, 23–24
Stephan, Klaas 17–18
stress, coping with 72–73
strokes
    brain damage 10
    language loss (aphasia) 17
superconducting quantum interference
      devices (SQUIDs) 24–25
Sutherland, Earl Wilbur 28
sweat, smell of 135–136
sympathetic nervous system 34
synchronisation of neuronal activity 25

Tass, Peter 24, 25
Taylor, Shelley 72–73
temporal lobe of the brain 10
testosterone 54, 59
    onset of puberty 63–69
thoughts, measuring 17–19
thyroid gland 52–53
thyroxine 53, 59
tibolone 95
tiririca, aphrodisiac properties 150
tocopherol (vitamin E), aphrodisiac
      properties 151
tri-iodothyronine 53
tryptophane 82, 83

uterus 61

vardenafil 156
vasopressin 59, 64, 69
    partnership and fidelity 76–77

vegetative (autonomous) nervous system 33–34
vertebral channel 12
viagra (sildenafil) 97–98, 154
    competitor drugs 156
    mode of action 155–163
Vincent, Jean-Didier 101
vinegar 89
Virag, Ronald 146
vision 21–23
visual cortex 9, 12
vitamin E, aphrodisiac properties 151
Vogt, Cecile and Oskar 5
vomero-nasal organ (VNO) 134–135
von Goethe, Johann Wolfgang 125

Watson, James 46
Weinzierl, Erika 81
Wernicke, Carl 5
Weyand, Ingo 132–133
Whybrow, Peter 53–54
Wisiniewski, Janusz 86
Witt, Diane 70, 71
women
    see also menopause
    attractiveness of pheromones 135–136
    characteristics attractive to men 40, 117, 128–129
    characteristics developed at puberty 62
    contraceptives 89–90
    coping with stress 72–73
    effects of alcohol 151
    effects of oxytocin 72–73
    effects of sex hormones 61
    eye contact 37–38
    flirting 38–39
    pheromone production 137–138
    tolerance of men's behaviour 75–76
Wurtman, Richard 83

X chromosome 48

Y chromosome 48
yohimbine, aphrodisiac properties 152
Young, Larry 77

Ziles, Karl 5